Technical Guide for
RainMaker Device
Ghost Consciousness
Catching Device
Zero Point Energy
Ascension Machine
and
Over Unity Coverage

Technical Guide for RainMaker Device, Ghost Consciousness Catching Device, Zero Point Energy, Ascension Machine and Over Unity Coverage

By
Kosol Ouch, Koeun Noun Ouch, David Lowrance, Martin Pott, Jerry Evans II and Vince Panella

E-BookTime LLC
Montgomery Alabama

Technical Guide for RainMaker Device, Ghost
Consciousness Catching Device, Zero Point Energy,
Ascension Machine and Over Unity Coverage

Copyright © 2007 by Kosol Ouch, Koeun Noun Ouch,
David Lowrance, Martin Pott, Jerry Evans II and Vince Panella

Library of Congress Control Number: 2007905120

ISBN: 978-1-59824-675-9

First Edition
Published July 2007
E-BookTime, LLC
6598 Pumpkin Road
Montgomery, AL 36108
www.e-booktime.com

Dedication

Dedicated to Carlos Sanchez, Jessy Sanchez, Jake, Tim, Jerry Evans II, Ben, Koeun Noun Ouch, Heng Vun Ouch, Nancy Rubby Ouch, Cham Nam Ouch, as well the many others involved in making all these books and projects possible. May God bless them all.

Natural Relativity

Preface

I thank all that have been involved in this quest to merge the Spiritual knowledge with the Physical sciences. There are a great many who have shared freely their insights and comprehensions. Each one brings mankind to a higher perception through the sharing. The magnetism site has been a "developing learning site" rather then a "teaching site," so may not be offered with the greatest accuracy if considered in sequence. The newer documents reflect embracing the models that have held up to direct experimentation and observed phenomena.

Since the collective development of the density sphere there have been many new discoveries for application, and the density sphere could probably have its own book. Kosol reminds me that none of us personally owns this knowledge and it is intended for all mankind, which I whole heartedly agree. It must not be sealed or hidden for personal gain of any one individual to the detriment of mankind. Kosol has reported that his new Density Sphere version resembling a star, can produce conscious Orbs. Even for a non believer! He is using much more wire then the original constructed by me some time back. Some 100 feet and adds magnets as well. A voltage of 10 volts at a frequency of 1.7 Mhz is reported to be a minimum requirement for this, and I would confirm this was where the RainMaker began to produce the Orb phenomena I experienced, and a much stronger personal interaction with the "Light."

Martin has now wrapped the crystal spheres with fiber optic light cables and has begun a whole new phase of conscious investigation. I too have confirmed this with 1310 nm Lazar transmission systems through scalar wound fiber cables. The first perception was the familiar "pressure presence" of scalar coils and tempic field acceleration. My new perception is that most of the physical world we see around us is truly powered by "Light pressure." In physics we call this "inertial momentum" and all the atomic particles appear to exhibit this quality. All the forces can be traced back to Light speed velocities at some point moving inwards, and it is here in the "Light" where all motion originates. On touching this I immediately knew, the "The Light itself" wants us to know of its function as Source powering the physical world. This is the one piece missing from the current physics, at best referred to as the "A vector potential", but I recognize this force as the linear distance force having no curl. It is not a new force but the original one, or the prime force powering everything.

These discoveries were viewed by myself as "Spiritual initiations" and have become a part of my current belief system.

Rather then become lost in the "Light" and the discovery of the Light connection, I have chosen to turn back towards the place where the conscious meets the physical to stand on the edge and look both directions.

Time

Time is the least understood phenomena in our existence. Time is a conscious process and requires memory on some level to operate. In the physical there is only motion, and yet we as beings are aware of time. Time can be pictured as an arrow. At the tip of this arrow is the present moment or the "now." All the physical world is contained at this tip and all field forces, but consciousness is not restricted to it. The physical mind lags the tip, generating mental images and memory bits to form a model of our reality. The model in our heads is not the reality but an image of it. As a meditator crosses forwards through the tip of the arrow it is discovered that leading the tip of the arrow is the Mental plane of Light. All that is happening here is being staged and created, and we as beings of Light are far more active in this process then we

10

are normally aware. One will discover that time does not stop at the tip of the arrow to become God awareness of all things in a single instant, but is functioning both ahead of the "now" and behind it. Time is a far greater concept then space or the field forces. Equations that try to contain both concepts will fail to work the first time that consciousness works to alter physical space. The structure is a hierarchy, space is a small subset of time. We as conscious beings, at our current level, exist above space, yet within time.

Time is rooted in the first split of the Yin and Yang acting to slow the time flow rate so consciousness can experience all possible conflicting dance. Space is the creation of consciousness acting along one density where time flow rate is a normal band of Light velocities. Consciousness turns inwards and projects space.

Within our consciousness is the ability to move where the "field forces" do not have any power, either ahead of or lagging the "now." It is in the smaller area of the "now" that all the physical forces operate. So with the manipulation of the physical world we are concerned with comprehension of the field forces operating through "motion" or the "tempic field force."

We as conscious beings do not operate outside time, but are rather vibrations contained within it. This field spans far more than merely the physical world of space and motion within space. Otis Carrs ship reveals the natural connection existing between space and time, showing where we as conscious beings actually fit. The ship was physically moved through space in an instant freed form the physical forces of tempic field, but the events were still moving through time. From the recognized natural hierarchy, space becomes far easier to manipulate then time. When one succeeds at consciously manipulating objects in space, then all the formulas attempting to relate time to space or forces within space become nonsense. Time is not a "field force" but a conscious experience. "Time" and "tempic field" are of different natures and follow different laws.

11

Identification of the Tempic Field [Review]

The nature of the tempic field is absent in conventional physics, [time flow rate relating to a field force that is alterable]. Wilbert Smith presented the model x,y,z, then came T E M. He lists the physical dimensions first [x,y,z] then T E M [Tempic - Electric - Magnetic] for the field forces. He makes reference to the nature of the tempic field as being consciousness focusing to produce spin. Once spin is present the Electric and Magnetic fields appear in quadrature. The impact of this model places "motion" or the force behind motion as the first field force with the highest field density [linear function]. This falls directly into quantum physics as what is now referred to as inertial momentum. All the atoms particles possess this quality and appear to spin without end, as sure as a light beam will travel straight without running out of power and stopping. Smith makes the first reference to this connection [from consciousness to motive power of the universe] that I have personally found, and presents it as a scientific formula.

From this presentation of the field forces, and direct observation of the physical world, I have concluded that light is the motive force powering the universe. In matter it spins, in energy it is radiant and propagates through space, but the one thing it never does is stop.

Light speed is the absolute limiting factor for all motion, but is also the Source powering it. It therefore appears to be the zero point field. It is not fixed in velocity or in momentum as matter. The indication that Smith suggests it is a field force, gives us an incredible new perception.

Tempic field is the prime field force. The first one encountered from the path of consciousness moving outwards. Certainly a major concept for a ZPE model.

Tempic field theory thus allows us to manipulate a third force, not merely EM. The consensus in particle physics is that all energy from the strong force area at the nucleus outwards is EM so all focus has been placed here and little on the propagation of inertial effects from the inner atom outwards.

The spinning Copper cylinder experiment shows us an inertial effect, but the first explanation from those presented with it has always sought to explain it as a magnetic effect. The result of eddy currents and reversed magnetic fields. When one looks closely at the actually inertial effects it becomes an interesting puzzle as to how the atoms spinning mass, which lies at the nucleus some 3600 times heavier then the electron shell, is projecting a force outwards to the magnet that would seem to be transferring the prime force of nature that distance. I have developed my own models on how this is possible but it must be recognized that the AG elements posses this quality to the extreme as compared to the other elements. They are tapped into the prime force of nature. This is the one with the highest field density.

Natural Relativity

This perception does not originate in science but in Spirituality and the "nature" of Spiritual perception applied to science. This is why it is presented separately from the above, but will allow a further step not presented above in the merging. OU. I am given the knowledge that "The Light wishes to become recognized for what it is." This is a direct quote and does not come from me but from it.

Consciousness and time originate in the spiritual and not in the physical. All matter is conscious. The nature of the Aether, the Light realm, is Light and Gravity. The Light is consciousness itself. Gravity is the attraction between awareness at the lowest level. Gravity and inertial momentum can be released from the conscious side of reality, this is shown in party levitation. There is a conscious time effect experienced by the weightless person.

Matter is built on consciousness units or quantum appearances of energy at the smallest level of physical appearance, and winks on and off as it moves through our density. Quantum physics has now reached this level of awareness bridging consciousness with the physical world.

The direction and velocity of a consciousness unit [CU] is it's vector and scalar. Its scalar represents its inertial momentum, but is regulated by its velocity which is not fixed. Its' light speed alteration factor is not relative to itself but to all light around it and the interaction between all that are present, which can alter how far it lies from Source as matter, its scalar.

The fabric that becomes filled with conscious Light at one local region from source creates the Tempic fields parameters. It is a field where velocity is relatively constant and defines the root speed of light as well as all motion we perceive in the physical. The underlying root principle of all motion is light speed, this is its' source and its' limitation for a given density. CU scalar and vectors thus all interact through a method of Natural Relativity. The two parameters we are concerned with opening to human consciousness are the ones involving time flow rate, or Tempic field, and Gravity.

The three forces appearing in the physical are the Tempic field, Electric field, and Magnetic field. Natural Relativity involves all three, however not the same. The only one that offers a reversed interaction is the tempic field because it directly originates on the conscious planes. If we set two conscious focus's against one another we get a higher energy output then if we set two magnetic fields against one another, or two electric fields.

As two CU's approach one another at light speed velocity they alter one another's velocity, not dropping it but raising it. This is happening at the light speed levels so is not perceptible at the physical until we learn to observe using our Spiritual nature, or until we see an object disappear or speed off at a seemingly impossible velocity. As the CU becomes matter, energy moves into the magnetic and electric fields allowing it to interact with other matter through the physical forces, but the root tempic field energy is always present and is enveloping it. Tempic field force falls off as a function of distance and thus its field density is greater but it does not cancel anymore then we can stop a beam of light from shooting through space. This is why the magnets in the spinning cylinder experiment are projected at such high velocities beyond what a magnetic field could normally do.

Thus as we set the Tempic field into opposition we see a density effect, or an acceleration of the time flow rate and a closer grouping of awareness observing more of itself. Light speed velocity raises. This increases the motional velocity of matter as well, altering its spin velocity and effecting all motional interactions. This raises the possible velocity of matter through space without loosing coherence of spin. We can move faster, and this is the basis for all UFO observed phenomenon.

The other two forces interact oppositely to this. When we set them into opposition they experience a cancellation, and their local field forces drop.

Two canceling magnetic fields, lower the resultant magnetic field present. Two opposite Electric fields, drop the radiated Electric field gradient. The energy between two opposing Electric or Magnetic vectors is not lost but is moved back into the Tempic field in the states of near light speed motion. Mass particle inertial spin. These inertial spin energies have both scalar and vector potentials that normally oppose, and move through all mater as two opposing chains of spin identified in the isotope lines, switching back and forth through the precession angles.

Energy can be bounced back and forth between the EM fields and the Tempic field, and the study of the scalar coils will show how this works. However to actually raise density we must isolate and then set into opposition the Tempic field. This is the only way to move beyond the conservation of energy. Thus opposing magnets in themselves do not result in an over unity in the physical, this causes only a shift of energy into the tempic field. A machine that will maintain a constant opposing tempic field throughout while manipulating energy between the Electric and Magnetic forces may be one solution. The Tempic Field Amplifier is one such possibility and seems to resemble the pictures of the Hubbard device, as well as the elements of Daniel McFarland Cooks' battery shown on patent 119,825 on October 10 1871. What was never disclosed in these examples was a parameter of Tempic field operation we discovered in the scalar or torsion coils, that of turns direction interaction. This is

15

discovered to have a direct link to the mass torsion or tempic field effects in copper.

Field Manipulation

The building blocks of the magnetic field are structured such that we can break them down in stages using opposition to result in three states of field forces. Magnetic which has all three vectors T E M. Electric having only two vectors T E, and Tempic fields having little or no appreciable EM. T [Tempic] always appears in two opposing flows with varying balance, and you can never isolate only one side completely so must always deal with two sides and a balance between them. Motion will never be canceled or become zero as voltage and current will. Tempic coils start to look more complex because they have dual wound cores in opposition, and 4 wires coming out of them rather then two.

Magnetic Fields in Opposition

The basic scalar canceling coils set into opposition both magnetic and electric vectors. These are the root structure of the magnetic field existing beyond the prime tempic field. A modulation to the input fields will move through copper where it cannot be detected using any EM equipment. The energy is transferred into the tempic field as inertial momentum alterations of the particles mass, and propagates Copper freely. At the other end of a long Copper tube these can be shifted back into the EM layers using coils at 90 degrees to one another. This concept is shown in the density sphere as well where a scalar canceling input coil can power a light bulb through the two other spin planes lying at 90 degrees to it, each is connected in a reversed series configuration and power is transferred through a scalar cancelling coil. This uses 6 coils in three planes, two in each plane.

Electric Fields in Opposition

As two large copper plates are set at 20 feet apart and each charged oppositely we see a voltage gradient reaching out as a function of distance squared. As the two plates are brought together to say 1/16" separation we see the force move extremely close together and the radiant outer field density drops drastically. Sensing equipment held a couple inches away no longer feel the radiant E field, as the presence of both charges very close and in balance cancels. E fields reach or radiant field density is reduced in size with capacitors.

Tempic Fields in Opposition

Torsion coils demonstrate the effects of setting torsion into opposition. The density sphere has shown us that we can easily create torsion forces that feel as though they will push the top of your head off by setting them into opposition. The consensus is that these fields do not cancel but accelerate in velocity as EM is forced back into the prime tempic field of motion. This pushes on Density itself or the zero point field. This is a conscious interacting field force as well. Here we discovered the importance of CW or CCW turns on a coil and how the tempic field is altered differently in each one. We can now split the torsions two natural opposing flows, then reverse one and recombine them back into one flow, simply by reversing the output phases. This takes 4 coils in the center of a system, and Sweets config is one application.

Tempic Electric [TE] Fields

Scalar coils can be constructed [TE coils] such that they cancel only the magnetic field and do not effect the Electric field by using reversed windings on each layer [CW - CCW]. The magnetic fields will cancel but the Electric fields will not cancel because both electric polarities will match on each end of the coil, even though the magnetic fields will be reversed due to wind direction. The field that becomes radiant will consist of a Tempic and an Electric vector. This is the model of gravity as well having both Tempic and Electric vectors, but absent of a magnetic vector. This coil design is explored in the TFA.

Establishing the Parameters of Torsion Coherence in Copper Coils

This was first observed as an interesting voltage gain effect but has now been recognized as an important tempic field effect because it allows the manipulation of energy to move from tempic field to electric field, thus can be used as a method of tapping the tempic field once it has been expanded through opposition. It is a study of coils setting at 90 degrees to one another, and how they interact along the three vector forces, magnetic, electric, and spin of mass or Tempic field.

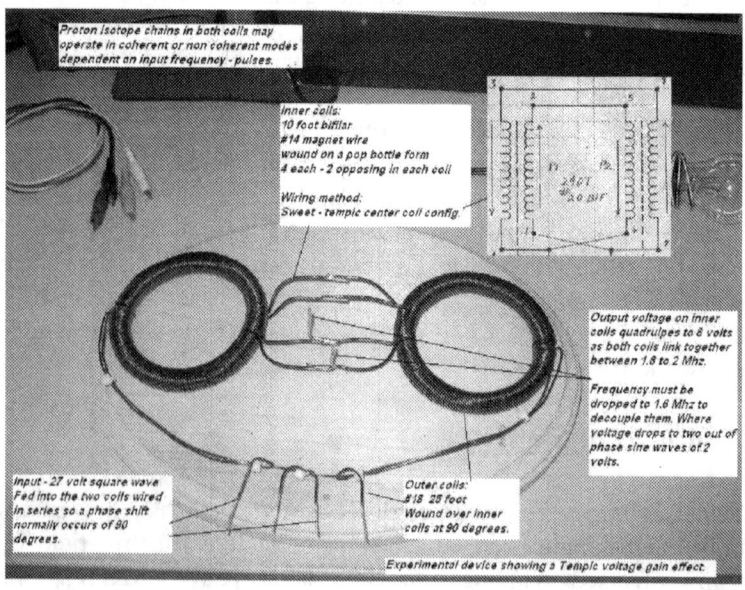

Proton isotope chains in both coils may operate in coherent or non coherent modes dependent on input frequency - pulses.

Inner coils:
10 foot bifilar
#14 magnet wire
wound on a pop bottle form
4 each - 2 opposing in each coil

Wiring method:
Sweet - tempic center coil config.

Output voltage on inner coils quadruples to 8 volts as both coils link together between 1.8 to 2 Mhz.

Frequency must be dropped to 1.6 Mhz to decouple them. Where voltage drops to two out of phase sine waves of 2 volts.

Input - 27 volt square wave
Fed into the two coils wired in series so a phase shift normally occurs of 90 degrees.

Outer coils:
#18 28 foot
Wound over inner coils at 90 degrees.

Experimental device showing a Tempic voltage gain effect.

Phase locking

Inner coils are scalar canceling and the configuration comes directly from the Sweet VTA as seen on the circuit diagram insert. Outer coils are torsion wound opposing from our studies of Torsion fields. One is CW the other is CCW.

As one slowly alters the frequency of the input signal in the above experiment there is a band of frequencies where we see a 4x voltage gain in the inner coil network where a voltage should not appear by current standards of EM theory. To understand how a voltage can appear on a 90 degree coil one must study NMR theory and look at it from two reference angles. The operation is a spherical model and voltage may be tapped from any angle, if motion of the field is present. The voltage appears between the upper solder joints and the lower solder joints of the coils pictured, not where we expected it across the lower two joints where Sweet is tapping the cold electric energy. The device did not produce the cold electricity in this mode of operation.

The voltage appearing is the direct result of the two 90 degree phased input signals on the outer coils pulling into phase on the inner coils and establishing a phase lock.

As the system hits this phase locking range the two phases that were cycling on the inner coils merges into one phase, but rather then a 2x voltage gain we actually see a 4x voltage gain. This is probably due to the magnetic vector canceling and energy being pushed into the tempic field then out into the electric field from two alternate mass spin directions turned into one.

A current flowing in a wire will effect the two natural opposing torsion flows setting at 90 degrees, a scalar coil can be used to split or separate them, and then recombine them by reversing phasing one flow to aid the other. As a magnetic field passes through a scalar opposing coil at 90 degrees, it splits the torsion flows into opposite flowing directions in each coil, we then reverse the polarity on the output side to recombine them in phase.

This effect corresponds to the two copper masses on each inside coil network moving into a synchronous isotope chain spin and beginning to establish a coherent torsion field throughout. There is a definite attraction of the Proton mass in each coil to begin to spin their mass together in time, and this is one quality found in Copper making it an AG metal. The inner coils were wound with heavier wire [greater copper mass] so this effect can be observed over a greater band width, from 1.8 to 2 MHz. To drop the phase lock one must tune off frequency all the way back down to 1.6 MHz where the voltages again separate into two waves at 2 volts.

In this experiment there is no over unity observed, the coils sizes are not well balanced, but the effect of phased locking is shown to be real, and linked to the spin of copper mass attracting into a coherent state. This can be used to reassemble a pure sine wave from a composite of energies at various supportive frequencies, and one of the most surprising discoveries is that the outputs from these kind of 90 degree coils is often a beautiful sine wave while the pulses going into the coils may be the worst of pulsing square waves.

Using the Density sphere with 44.5 foot scalar coils at 90 degrees we see a voltage gain jumping from 27 volts to about 100 volts using this method of phase locking between scalar windings at 90 degrees. All the coils are identical in this case in both mass and turns. This system shows us only one side of the system gain, voltage. To generate actual power we will need to add the current gain side as well found in the Tempic field amplifier. This requires a scalar voltage canceling effect to produce a magnetic field gain as well and then reassemble them in a series configuration.

The Tempic Field Amplifier

It is hoped that using TE coils we can work with E fields to capture OU. The E field in motion, having no magnetic field, has no inductance. TE scalar coils are unique, they cancel only the magnetic field. The TFA platform is set up to experiment with many possible combinations of coil winding methods and includes a copper density sphere. It is based on the Hubbard photos at the base offering 8 independent TE coils. It becomes a platform for mapping the field densities of the field forces, where now we have three to deal with rather then only the Two offered by conventional models.

You can reverse any two vectors of the field forces and the third is not effected.

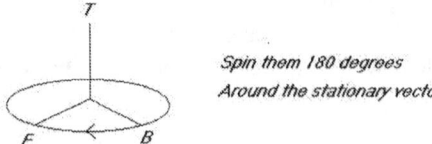

Spin them 180 degrees
Around the stationary vector

In a linear coil of wire the E field will move down the wires length all one direction and thus can be dealt with as one E field.

The voltage will spread along the coil and create a distance squared force around it with a gradient along the coils length. Also a small gradient on each turn.

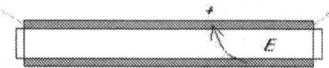

The magnetic field is a function of current in the wires loops on each turn.
The magnetic field will apear running lengthwise down the coil with poles at the ends.

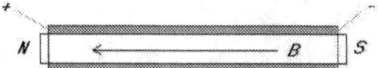

If we reverse the electric field, the magnetic field will flip also - referencing the top of the page we see that the tempic field will not be altered in this reversal. The B field and the E field will both be reversed.

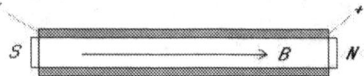

The Tempic field follows the wind direction of the coil as we discovered with the large torsion coils. CW = female or lower torsion state CCW = male or higher torsion state.

A coil with both winds - one on each layer will establish a tempic field gradient opposing between the layers - This field will set 90 degrees to the magnetic and electric vectors.

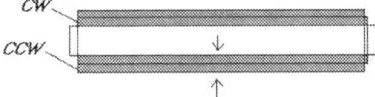

If we now feed voltage into the coils the same direction - voltage gradient along the length of the coil will be aiding and not cancel, magnetic field will be opposing and will cancel - tempic field will be opposing because wind direction is reversed.

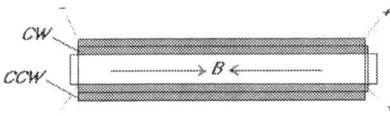

If we now reverse the polarity of one coil the magnetic B field will now move into an aiding position and the E field will be opposing along the length of the coil. The Tempic field will still be opposing as we have not altered the turns direction.

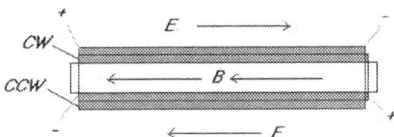

The longer the coils the more voltage gradient cancelling we will see end to end. These two different scalar cancelling effects will toggle between the fields.

Diagrams showing the two possible ways to power the TE coils.

Tempic Field Amplifier Coils

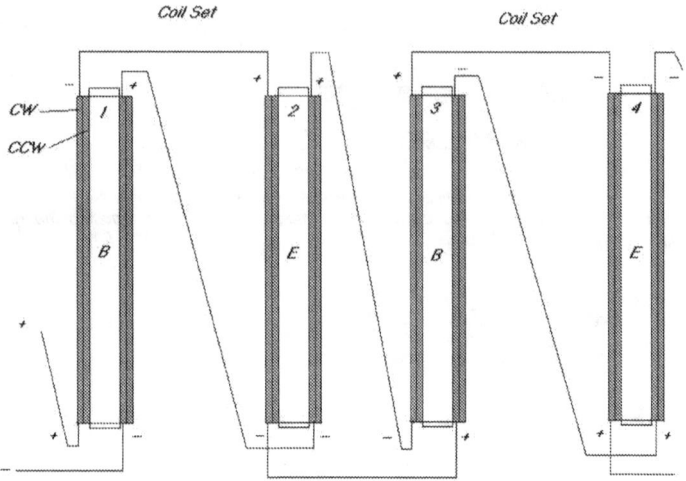

B = Coils that increase the Magnetic field
E = Coils that increase the Electric Field
Coil Pairs must achieve a phase lock to transfer torsion into the EM layer

Coil 1 to coil 2 - Reverses E field and B fileld - tempic field remains constant in turns CW CCW
Coil 2 to coil 3 - Reverses CW to CCW in each copper loop causing a breaking of the isotope
    chains - This breaks the tempic field into opposing spins and tends to allow phase locking
    in coils 1 and 2 which cause an energy gain in both the magnetic and electric vectors in the
    coils that aid these fields.

4 coil sets are needed to close the loop as a 90 degree phase shift is expected to apear
between each coil set.

                                     Dave L   c_s_s_p   4/14/2007

Tempic Field Amplifier - 3

*Longer coils and higher voltages should gain more EM field increases.
Spacing will become criticle as coils too close will have an E field interaction.
If wires between sets are made a bit longer the torsion breaking will be accomplished
easier. Wires between coils in the same set should be a short as possible.*

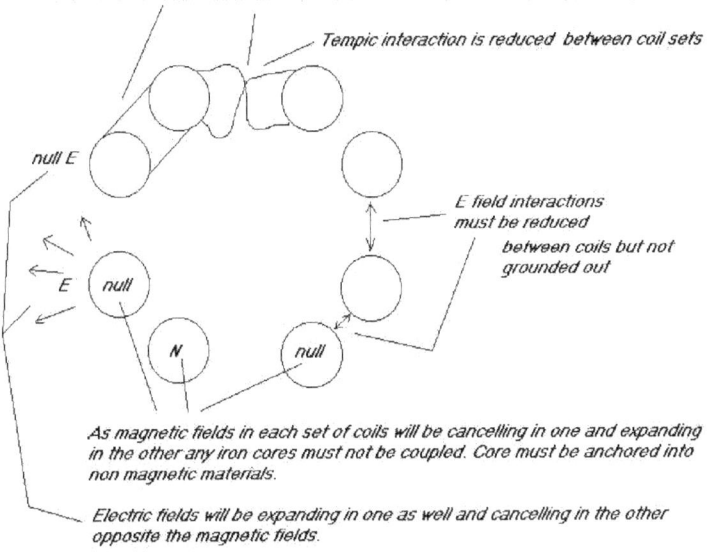

Tempic interaction is reduced between coil sets

null E

*E field interactions
must be reduced*

*between coils but not
grounded out*

E

null

N

null

null

*As magnetic fields in each set of coils will be cancelling in one and expanding
in the other any iron cores must not be coupled. Core must be anchored into
non magnetic materials.*

*Electric fields will be expanding in one as well and cancelling in the other
opposite the magnetic fields.*

It may be possible to place iron cores only in the B coils and capacitors across the E
coils such that the device is tunned to its tempic phase locked frequency as it is
powered up and very little input energy may be necessary for startup.

A breaker or two breakers can be placed in the copper loops to ensure shut down on
over current and runaway conditions.

Dave L c_s_s_p

4 / 14 / 2007

TE coils

TE Coils in the Tempic Field Amplifier were labeled in two
forms. B coils and E coils. Each operating on the three field
forces differently and can be constructed in many ways. They are

basically the same coil, one voltage polarity reversed in each resulting in a different vector scalar interaction.

B coils

In the B coils the magnetic field is found to be not canceling but supportive. As both coils have opposite turn direction, when we place current passing through them in opposite directions end to end the magnetic fields generated are supportive. The voltage gradients however appearing along the wires end to end become reversed and a scalar vector canceling of the voltage gradient is present, and the ends act like capacitors withdrawing the charge density. This pushes electric energy into the Tempic field increasing it. Since both windings set in opposition along the Tempic field vector this energy is first expanded in velocity and then hopefully pushed into the magnetic field.

E coils

In the E coils the magnetic field is found to be opposing or canceling. The voltages are applied such that current flows the same direction along the coils and a large voltage gradient becomes present across the coils with a greater field reach. The Tempic field is still in opposition along the coils so the energy moving between the canceled magnetic and the electric fields should experience a gain since it all comes from the Prime Tempic field which is now expanded.

As the energy moves between B coils and E coils wired in series it should experience a Tempic field gain as well as a 90 degree phase shift.

Where ever energy moves back out into the E or B fields it should reflect this tempic field gain present in all the coils.

Coil Layout

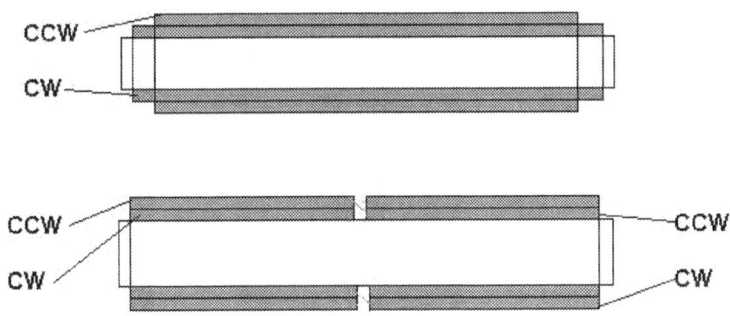

To balance the copper mass of each side of the tempic field and allow for smooth phase locking the coils should have the same wire length and gauge on both sides of the system. The two spinning masses will then easily come into synchronization, both spinning opposite directions away from the center of natural tempic operation.

The coils can be wound longer on the inner layer as pictured in the upper illustration. The outer layer can also be separated at the center so that the end section voltage gradients better cancel. On the lower diagram the coils are wound in four sections then crossed at the center so that each winding will have the same wire length end to end. This may allow both to be wound simultaneously approaching one another and then overlapping at the center and continued to the ends. A small rubber washer or ring can be placed at the center point offering a place to hold coil tension and offer a reversal point for the coils wires to each flip. On the 3/4 inch Aluminum cores pictured below rubber hose washers work well.

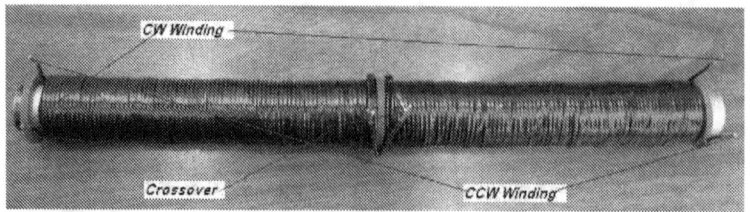

In no case however should a coil change its CW or CCW turn at the center point, this will reverse its tempic field direction. The lower example shown will produce two coils with the exact same copper wire length, and voltage gradients appearing evenly along them near the ends. On this coil construction it is recommended that the leads be marked clearly as it will be very hard to identity them if the coils are flipped over. My first attempt shown in the photo is 1 foot long, with two sections of #14 using 78 turns on each side that take up 5.5 inches. Two layers, inner layer moves to outer layer at the center.

The B coils can be wound on iron cores to increase magnetic field and the E coils on plastic or dielectric cores to increase the E field. A layer of electrical tape may be considered on the coil form and between the layers on the E coils but will probably not be helpful on the B coils. Experiment is in order and first attempts use 8 each all identical coils.

References:
David Lowrance, Author, Designer of the TFA platform
Dell Coleman, Tempic Field Theory of Wilbert Smith
Andrew Bellon, Concepts of time
Kosol Ouch, Spiritual spherical systems offering conscious interaction
c_s_s_p group [The collective progression of comprehension]
4 - 14 - 2007

ZPE Coils

Preface

Intended for the betterment of mankind, these coils are the result of many working in the alternate energy research field of public domain information. This is a collective project shared by many sincere and giving individuals, with the intention of discovering a new source of power and healing for our planet.

Definitions

Lathem coil - a coil based on a special length of wire [44 feet 6 inches] found to resonate a free torsion energy that has been discovered to be present in many places on the earth. They are typically wound into alternating scalar canceling and normal clockwise or counter clockwise forms, and wired in series to produce healing devices that are self powering to a mass of crystals present with them.

Terra coils - a coil based on the long side of a triangle distance crossing four Terra lines at opposite corners. The distance can be found experimentally by "pruning" but is theoretically about 45 feet 3 inches / fractions or multiples of this length to capture wavelengths or harmonics. As earth points may vary. Terra lines may be palmed or sensed in a local setting and various lengths can be connected to the RainMaker coils to sense the power levels of the captured fields.

Kosol coil configuration - One scalar bismuth coil is parallel wired to one normal wound coil with an iron content and both are setting in the same alignment. This system provides a strong linear torsional powering system. If placed into platonic form can create a density sphere system.

Limax Coil - Coil developed by Martin Pott, is wound with a floppy disc type flat ribbon cable. Ends are then spliced alternating to achieve a reverse current flow in each wire moving up the coil. This coil can be opened with a switch to shut it down. Coils can be now wrapped quickly to almost any depth.

Weave Coil - Designed by David Lowrance based on the description of Wilbert Smiths writings, a coil with each single loop reversed from the one next to it. Loops are placed on the coil form then twisted around one another at each of 180 degree points along the sides of the coil moving upwards. Normally a one layer coil is plenty for a bismuth core to become notably strong using a function generator. This is the coil detailed in the RainMaker vortex generator.

Tempic to electric conversion coil- Two coils of normal wind in non resonant lengths can be set up at 90 degrees inside a Faraday cage and tuned to receive torsion field waves induced using mobius or other scalar canceling coils. The coils must intercept the torsion waves at specific angles as torsion waves are polarized much like light or photon waves. Measuring equipment can then be attached via coaxial cables to monitor the presence of torsion induced waves. It was discovered during tests that many electronic devices actually radiate fairly strong 60 cycle torsion waves not detectable by normal EM equipment, and this is a common practice used to cancel unwanted EM.

Platonic form - Grouping of magnets or Torsion coils such that all torsion fields are equalized in all three dimensions forming a smooth confinement of space in which a higher density may be contained. Searl disc is patterned as a 2 dimensional platonic form. Kosol spheres as three dimensional platonic form. RainMaker base is a 2 dimensional form and the spherical crystal on top pulls this into a three dimensional one. Most all tube devices are single dimension.

Linear forms - Any device creating a strong torsion field in only one dimension, must be balanced with "mass of crystal" to avoid possible torsion sheers. This includes most all tube devices as determined by direct experiment.

Clockwise coil - A coil wound such that energy will move through it in a clockwise rotation as seen from the energy looking forwards to where it is heading.

Counter clockwise coil - A coil wound such that energy will move through it in a counterclockwise rotation as seen from the energy looking forwards.

General

It is discovered that there are two ways to wind coils and both produce different atomic interactions as to Tempic, Electric, and Magnetic field forces.

To determine wind direction of a coil, sight down one end of the coil and pretend you are moving into the wire closest to you and moving away. If you end up turning clockwise then you have a clockwise coil. Note that if you flip the coil over and do this again you still have a clockwise coil. Both coil winds are unique in this respect.

A clockwise wind will favor Proton spin, as energy moves through it, Proton spin is added to, and Electron particle spin is lowered. It is expected that this would be reversed in a counter clockwise coil. Noteworthy also is if Electricity moves through both coils it will produce a magnetic field in each that is reversed due to the Right Hand rule for magnetism. Proton spin results in torsion fields and Electron spin results in magnetic fields, so each coil wind alters both forces differently.

Energy in Copper Coils

It is possible for all three forms of energy to move through copper.

>Tempic - Torsion fields from Proton magnetism
>Electric - Electron flow at electron shell
>Magnetic - Electron shell in motion

With Terra coils we are working with the Tempic field and not the Electric field. To produce electric flows, some means must be used to reestablish the EM field from the photonic or torsion interactions. Tempic fields do not cancel but expand when set against one another, and this is the means of manipulation used to our advantage.

A Torsion Capturing Coil

This coil is basically a dead coil as to perception however within it is stored, invisible, a resonant torsion force much like a battery stores electric energy. If used in conjunction with other similar coils at 90 degrees to it or with other scalar coils the Torsion can be manipulated and becomes strongly perceptible. Altering the coils "wind" will alter its Torsion output. It is a linear form so is not normally used alone.

This coil is constructed from a 1/4 inch bolt 2 inches long, 44.5 feet of Grey insulated 24 gauge copper wire and a short length of electrical tape to keep the final wraps tightly held in place. Wind the coil along the bolt thread such that it follows the wind and a clockwise coil will result. It is then carefully wound to many layers back and forth to complete the length.

When connected directly to RainMaker this coil powers the unit to head popping power levels and the crystal sphere becomes too intense for prolonged use, showing the actual Torsional potential of the coil. The RainMaker unit can be used as a sensing device using this method, to determine the torsional state of coils and coil combinations if one has developed a working connection to it. This requires that the weave wind bismuth coils be used or some other non resonant length coils in the base.

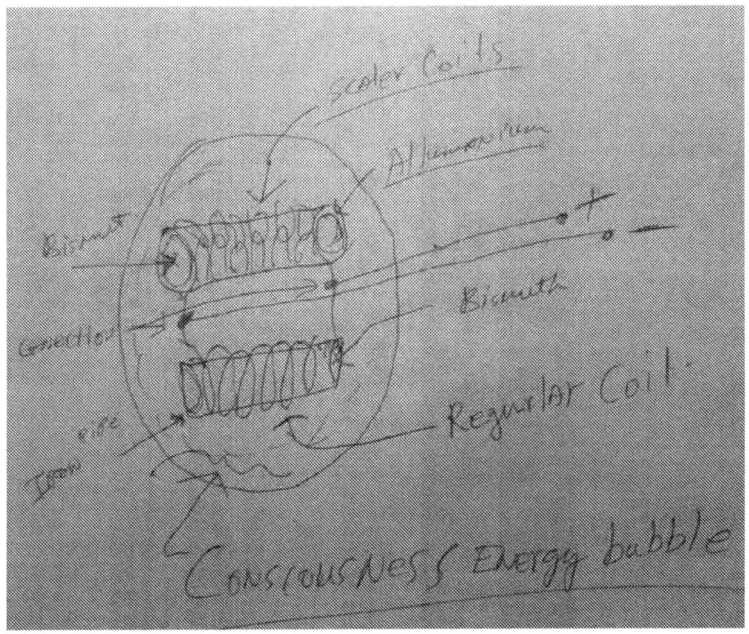

This combination as diagrammed by Kosol in parallel configuration, if placed into the inside of the RainMaker base will also produce very powerful Torsion fields, and in the case of my large 6 inch radius sphere are too powerful to leave operational for very long as they are in linear form.

A Simple Donut Coil

This coil will demonstrate to one the nature of the earth torsion fields present and allow for experiment on two dimensions of torsional field compression.

Construction is done using two torsion resonant lengths of twin lead 24 gauge wire. The first is wrapped around a 20 Oz plastic coke bottle. The bottle sides are now compressed to remove the coil and the second coil is wound around it to form a donut. This is done using clockwise wrapping techniques [check description above for clockwise coil]. Second coil can be wound from the center of the wire passed through the first coil so shorter ends will be pulled through on each wrap, also a clockwise wrap.

We now end up with 8 wire ends that can be configured to show the energy interactions. Inside is one black and one white horizontal wound coil and outside is one black and one white vertical donut shaped coil. The two dual coils interact at 90 degrees to one another and can create torsion fields in 2 dimensions.

To configure either coil as scalar canceling simply take opposite white wires and connect them to opposite black wires reversed, so the energy moving through the two coils passes opposite direction through each. Black to white in each case from opposite ends of the coil are shorted to form a scalar coil that can be shut down by opening the leads. Since the torsion resonant coils can be very strong this wiring technique saves one destroying the coils if things do not proceed well.

The most notable effect produced from this coil is not what one would expect. If the black and white twisted wire ends are connected together as a single coil in each case and not in scalar canceling mode, connecting one end of the horizontal coil to one end of the vertical coil produces a tempic field receiver. This is sensed as a feeling of "joy" by myself. If one now holds the other two ends in a hand touching skin, but not touching one another, the coil will begin to receive vibrations from any elements placed inside it. I have never used a Witness well or a stick pad, but this setup seems to offer a sensing effect for feeling the NMR rates of different forms of matter.

A Scalar Canceling Laythem Coil

A two dimensional coil form, if placed on a washer will focus most energy off its sides. If used in weak form will offer a standing torsion manipulation. It offers ease of winding and a quick access to torsion fields if control is not desired or necessary at low output.

The Toroidal coil shown above was altered in my designs to maximize power output by winding it on an iron tie wire coil form 28 feet 8 inches long and wound into the donut shape. [Lyle winds them on washers in the Star Link System].Then the Lathem coil is wound over this after folding the 44.5 foot of 18 gauge wire into four lengths of 11 feet. The 11 foot group of wires is pulled through the ring to its center position and then wound both directions. This speeds coil winding considerably for the toroidal shaped coils. 3 inches of excess wire then is left for connections [the red wires shown above].

The coils are quick to construct and powerful, but cannot be shut down easily, so expertise is required to balance the system using proper crystal to copper ratios.

The Density Sphere

Built by David Lowrance on 2/28/2007 and released to the public domain as no longer patentable.

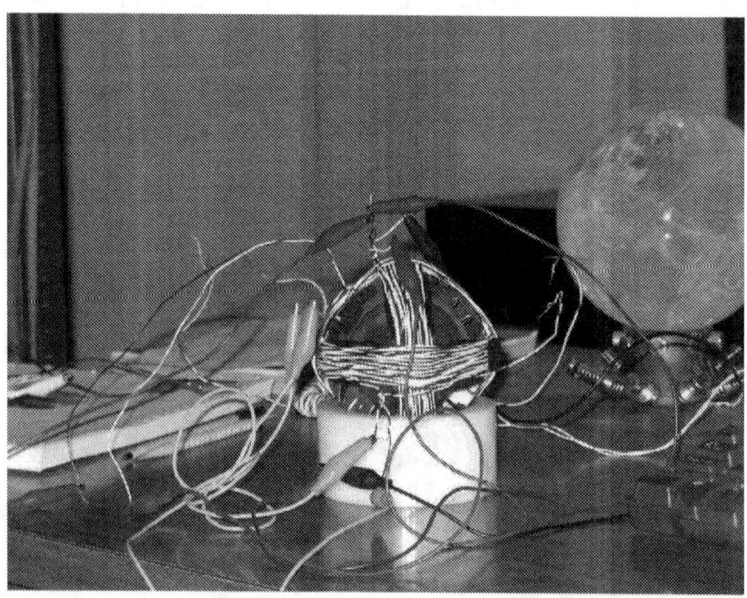

A full function density generator offering torsion on all three dimensions as pioneered by Kosol Ouch. This can be adjusted in power output by turning its Earth alignment. The three scalar canceling coils are Lathem length coils but offer a control for shutdown not present in the Star Link systems, that was offered in the RainMaker ferrite base units early on. So this generator can be used as a stand alone controllable device. It is suggested that other coils be worked with first to determine whether these coils lengths are adequate for your locality. Ideally you will have your own custom lengths determined before constructing a full Density Sphere, otherwise the results can not be guaranteed to be consistent.

This device can fully replace the function generator used on RainMaker units and should not produce any notable headaches of itself at high power levels. The strong fields may however be altered as they are used to power other devices, which can result in a linear compression, so care must be taken in design of the powered devices as well.

Construction

The forms for the wire are constructed by cutting cardboard rings and then taping these rings to form a full 3 dimensional coil winding container. Sphere chosen is 3" copper laced quartz having a rather strong field to begin with. Sheet rock knife, scissors, and various tapes are used. It is desired that the crystal be free to turn inside the unit later so care is taken in construction.

Rings must set on the sphere slightly to one side of center, but is not too critical as another ring is placed directly on the sphere to separate the rings.

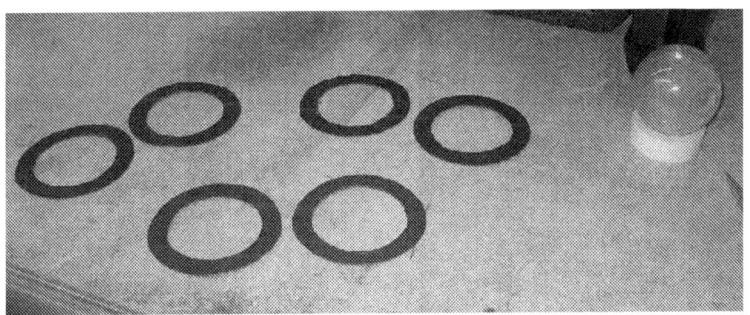

Showing the six rings necessary for all the materials to hold the coils on.

Fold the rings in half and trim so they loosely fit over the sphere on a quarter section.

41

Cut lengths of cardboard also now to seat under the rings. It must be custom sized to fit the coils wire amounts for width and sphere size. A coil can be wound ahead of time to estimate thickness. Coils can be any torsion resonant length or multiple resonant lengths for your local earth resonance. The strips are fitted then taped to cover the sphere on all three planes of spin such that the crystal may still be moved inside them for alignment if this is desired.

The rings are now cut into quarter sections.

Quarter sections are lined up on the crystal and folded to form into triangle shapes then taped into place using electrical tape.

No tape contacts the crystal surface but is contained on the coil form such that the crystal may be realigned inside the coils after construction.

Showing completed coil form around crystal sphere allowing 3 coil winding containment channels. The unit is set in the top of a tape roll in this case to keep it stable while tapping the last pieces into place.

More tape is added to seal all joints and provide a clear surface for wires.

Coils will now be wound in layers clockwise wind, one layer over the other in each quadrant until the complete wire lengths of twin lead 24 gauge 44.5 foot lengths are all used.

All three lengths must be started simultaneously, then the lowest quadrant is worked one at a time on each layer. The coils are wound by taping one end of all three wires in place at starting position and working one spin plane at a time for one layer. This will produce three interlaced coils of twin lead.

Connections

The coils are all now configured as scalar canceling by joining opposite ends of opposite colors on each one. Next the scalar coils are all three wired in series to provide a total copper interaction that flows through all of them. Two of these can be opened to be used as outputs to the RainMaker unit for direct sensing of the energy inside the Density Sphere.

Adjustment and Control

Align the sphere of the Density generator so one coil is horizontal and the other two are about 45 degrees from the North Geomagnetic pole of the earth.

As you turn the coil now along its horizontal surface the intensity of the field will be altered that is emitting from the RainMaker and can be felt at any angle coming from the RM sphere. Direct input can be sensed also by touching the Density Sphere with finger tips on its crystal surface.

Coils can be deactivated by opening all the opposing coil leads and disconnecting from the RM unit.

The energy produced from the Density Sphere is "fun" and would seem to contain a high frequency healing crystal component vibration. At present I would class it as angelic female energy.

Kosol Ouch, Koeun Noun Ouch, David Lowrance, Martin Pott,
Jerry Evans II and Vince Panella

Tapping the Earths Tempic Field Grid
[Theory 2 - 12 - 2007]

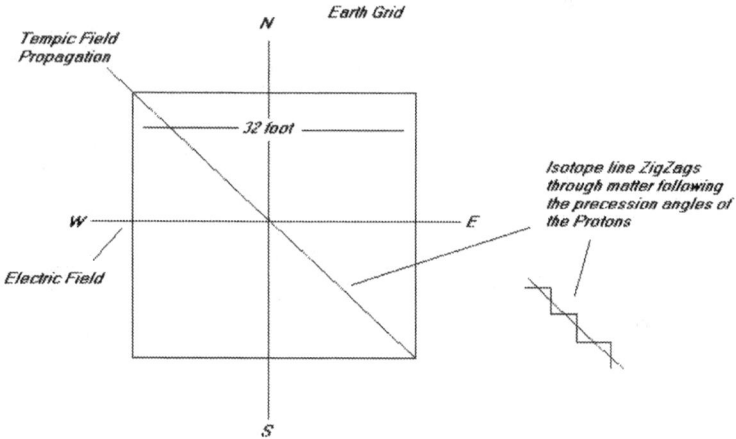

General Description

Torsion moves through matter along the Protons isotope line, switching back and forth in two directions. This sets up two torsion fields around the earth setting at roughly 90 degrees to one another. The tempic field flows at 45 degrees across the grid as a tremendously powerful torsion wave. The earth's grid has been mapped at the Oregon vortex, where lines appear at 16 feet apart across the interference zone, where tempic sheers are strong enough to alter weight and size of people standing inside it creating observable density effects.

Velocity of propagation was recorded by Tesla to be 471,240 Km/sec. But this is the resultant velocity point to point. The tempic field actually flows a greater distance then its propagation distance as it turns in each atom. The actual light speed is higher based on the precession angles of the Protons.

Experiment with the tube devices has shown us that a Torsion wave will propagate along a 45 degree line from a magnetic field. Thus placing a Copper wire on the earth grid along this angle will flow the earths torsion field through the copper. Length of this wire is calculated based on a 45 degree triangle with 32 foot sides. The hypotenuse of the triangle is 45.2548 feet. At this angle the wire should begin to resonate a standing tempic wave, or Tera wave.

Along the Tera wave, since it is the prime force, is the earths consciousness.

Tempic Waves on the Earth

[This is based on understanding of how the tempic wave operates, how light can be polarized, the two slit experiment, the earth grid, and Proton torsion operation.]

Tempic waves, I draw as a circle moving along its edge. This was my first model of light, and has never changed in many future models that continue to support it, even before I had the labels. It is moving along the shell of the conscious layer trapped in the density of its velocity. But its velocity is cycling, so this gives it a **width** along the density but not along the Magnetic or Electric vectors of vibration. It is physically moving back and forth from and to Source as it propagates along the control vector and the perception vector. Chasing its own tail playfully enjoying each turn either to or from Source, and feeling tremendous pure joy at each turn. This Joy is directly experienced at Light awareness, or bringing in the Light to the third eye center, which is very probably a Proton photon receiver.

Those EM vectors are there but very small, keeping the photon in the C velocity. But the Protons C is not the same velocity slightly higher then normal light and not so fixed, although its wavelength is longer at the Proton, its velocity is higher. We do not normally see Proton photons [being in the Mhz wave size] at 1.5 C we feel them, there wavelength is several meters, but a Copper antenna the correct length would actually flow this tempic wave as a longitudinal wave, because Copper will flow both electric- magnetic, and torsion fields. With a two loop system the EM can be injected into the wave and move back into the electron shell. **In the process the light energy slows down to C**. The math for antenna design should be almost obvious knowing the center velocity of the wave. The tempic receivers of copper will be much longer then EM receiving ones for the same frequency. This also speaks directly to the earths lay lines, and these waves may very well be the same as they are sensed rather then measured by any EM equipment. I know because I have tried at many points on the earth to measure them. They are sensed at third eye, but not with compass.

Here we have light which we know a lot about at velocity C, and then we have Proton torsion waves which we have never been able to see or measure at 1.5 C. They should operate the same however the Proton ones may be swinging much further in velocity on their loops.

On opposition. We observe in magnets that the tempic field is the one that does not cancel, but increases its velocity increasing field reach in space due to higher velocity. The electric and magnetic do cancel, however the tempic does not. Putting two tempic waves in polar alignment, and then 180 degrees out of phase will not make them smaller but only **increase** there velocity. They will disappear if they move through density. This is the model we see with Carrs spinning cones where the tempic or torsion components are moving against one another on the physical side and time flow rate increases. If the cones spin together rather then opposite directions they will not increase density but may even lower it, unknown at this point. If we are in the craft while this expansion of time happens then we go along, if we are outside then we stay behind. Caught at the edge matter is destroyed if a density barrier is crossed?

The photon is defined by Plank and has a finite energy value for light speed photons from the electron layer. It is in the nanometer size and a very high frequency much higher then their precession rate at the electron shell. Photons from the Proton layer may have a constant as well but only if they stay in our density. As they are canceled they would increase the velocity of their photon structure through space. Why the difference with electrons and protons in generating photons? The electron shell is the point where the negative energy loop starts and ends, the shell is where the outside world is kept balanced. The Proton layer swings through a tempic loop pulling on it, using mainly torsion forces. If the electron layer is stopped or canceled the proton layer then swings further then normal and tempic waves now radiate out of the atoms, and we **feel** them.

The scalar EM canceling coils are causing this interaction and the photons exchanged thus may be at a much higher density, while the coil is peaking in opposition at the electron layer. We now see variable density effects.

The earth has a crossing system in place with 16 foot centers I believe as well. These are most likely torsion waves. They manifest bent trees at the Oregon vortex, where the trees think they are growing up but tilt 90 degrees along these lines. Gravity is lower at the center of the vortex and people shift there size physically when reversing positions.

This is indicative of sever tempic field gradients present at that location. The earth has other points manifesting this tempic interference pattern. The lines are for some reason entering a canceling phase mode and altering density at these points on the earth. The Protons are able to follow these waves and so the atoms shift density.

I believe that while the torsion wave is altering its velocity on each cycle, the center point is found at the neutron in the diamagnetic field so matter will follow it. The coherent torsion field we all sit in, can be modulated with these torsion waves, to effect gravity and light speed. At the physical level, the effect is reversed however and tempic vectors opposing down here do not cancel but add to higher velocity on the conscious side. In scalar cancelling coils this is **amplitude modulation** at the electron canceling layer. The alteration will be a direct result of how hard we push the electrons shells in opposition, and a function only of amplitude of the waves placed at this point. To push density further, then we have been doing with our function generators, requires higher **currents** through the coils. The up side is that the currents will only be active for a few microseconds and then cancel.

If copper receives the tempic wave along a 45 degree angle of incidence as we have observed in the tube device as it propagates along it. Then a wave 16 foot long [earth grid] will need an antenna wire found using the formula for long side of a triangle. L = square root of [16 ^2 + 16^2] = 22.627 feet. Lyles wires are just about twice this length. His device exhibits the same unpredictable levels as Sweets, it will go dead at times and then become stronger at other times. Earths torsion field is shifting. Lyle is tapping into the earth at the Proton torsion layer.

We can build a tempic field receiver based on math at this point, using coils of correct length wires setting at 45 degrees. The EM coil will be shorter a radius of 1 and the tempic a radius of 1.5 should come very close.

The Star Link System

I have been studying the Star link as created by Marion Lyle Lathem. It shows us an ability to build a devices using no magnets only copper to power the unit based on 44.5 foot length of wires used all through it on 7 coils altering from normal to scalar up the base, then connected in series. The center shaft holds a string of crystals at a higher torsion column and circulates the energy through the copper wire loop similar to Sweets coils, and the torsion probe receiver design I came up with earlier. The crystals energy is immediately present and envelops an area around the units.

Photo shows my duplication of the coils of the Star Link system with an increased physical size. The wire is #14 up the tubes and funnels, each resonant to the Earth grid. The scalar coils are 18 gauge wrapped on 28 foot 8" rolled iron tie wires to form the cores. All the 7 coils are 44.5 feet long. With the coils opened placing a crystal in one funnel end, you can strongly feel their energy and a torsion vibration is set up in the system.

Experiment to Find Lengths Accurately

To accurately find the correct length of a tempic receiving wire for coil winding at ones location. Stretch out a length of Copper wire across the East West line of the earth, then move it to a 45 degree position of the Axis. Start at about 50 feet length, and slowly prune the wires ends until a torsion wave becomes strongly present in the wire. I would guess this may be palmed, or sensed in some way. Now fold the wire into quarter lengths and see if the torsion field is still present at this earth angle but expanding to a stronger level. This is now the form used to wrap the scalar coils, and the coil in the torsion receiver.

About 11 feet long it is wrapped around flat washers, but I would believe that once a good resonance point is discovered, this wrap could be incorporated in many units as a powering source.

As an alternate we can design a dual coil receiver to effectively cut these to earth resonance using a scope, with the probe located probably near the center of the wire at 45 degrees.

Copper Wire Length Calculations

Tera waves C = 471,240 Km/sec   EM waves = 300,000 Km/sec
45.2548 feet = 10.413Mhz   28.81 feet [Most probable
                                    correct size]
22.6274 feet = 20.826 Mhz   14.405 feet [based on 16 foot
                                    wave]
11.3137 feet = 41.652 Mhz   7.2025 feet [quarter wavelength
                                    for 32 foot earth grid]
    Based on a 32 foot earth grid these wire lengths should be close for some powerful resonance effects in tapping the earths tempic field. These could be called
Tera waves. Due to end effects and wire types the wire lengths may be slightly less for the EM coils.

Pushing Density Upwards With A 1/4 Wave Foldover Stub

Tesla measured the propagation speed of the earths Proton torsion fields at 471,240 Km/sec. This is the number he found from direct experiment.
    The Torsion field moves along the precession motions of the Protons and takes a 45 degree turn between each atom off the magnetic field or isotope line. This is actually a 90 degree turn for the torsion fields, so the torsion fields would form into a crossing grid, but propagating at 45 degrees to the grid. This is why the calculation is lower then actual velocity which should be up around C x phi I believe. The tempic field moves in a zigzag pattern through matter. But we know the end result from Tesla.
    Placing 4 each 1/4 wave wires along one another forms what is called a canceling stub. However we know that a canceling stub will cause an increase in velocity for the tempic waves and not a cancellation as with EM waves. If we use the full length of wire 45.2548 feet stretched out at 10.413 Mhz, then place a 28.81 foot wire along the East West magnetic pole joined at one corner, out

of this wire I would expect to see a fairly strong 10.413 Mhz EM wave in resonance and begin to radiate out into space.

Now if we fold the Tempic wire into quarter lengths, this will push the light speed constant up wards, and the wavelength will shorten a bit until the wire loses resonance. So here is where Lyle discovers that around 44 feet he has a density and resonant coupling to the earth that will naturally appear, but not always consistently. To move beyond this limitation we can place more sets of wires along this each one slightly shorter, and as the density factor moves upwards the resonance will move to the next wire set. In this manner we can build a coil that should maintain its power level over small shifts of the earth as well. Alternate between normal coils and scalar coils slowly becoming shorter from 45 feet down to how ever far one wishes to accelerate light speeds. Then at the chosen place we set up an Electric coil to tap the power. Frequency will increase with each step also starting around 10Mhz, and moving higher with each coil.

There may be other ways to do this also, but the wire lengths must be laid out to capture the field as it sets, then scalar wraps of 1/4 wave canceling to shift it upwards, and then capture it at the higher light speed we wish to operate with shorter coils. The effect will move, and the wavelength will shorten the harder we tap into it with 1/4 wave canceling antennas.

Now the Electron coil at 28.81 feet of wire is placed along the earths East West line at 45 degrees to the Tempic wires, I believe, this is also perpendicular to the earths magnetic field. It Can be tuned to a higher frequency also as the density factor moves upwards in the Tempic coils to find its stability point. A shorter wire can use a tuning coil or a cap for adjustment since this is normal EM at approx 10.413 Mhz and then moving upwards. We know that wiring the two wires now in series should produce electricity, if both lengths are tuned correctly there may be considerable power present for coils of the correct angles and lengths.

On Magnets

Now to magnets - if a magnet is brought anywhere near the system close enough to alter the isotope lines in the copper wires then distances all change, and operation shifts as precession angles are tightened, but only in the tempic wires. This causes a lengthening of the distance that the tempic wave will move down the wire. Proton precession frequencies rise with the function of the magnetic field they are setting inside, and the precession angles probably shift. Two possible outcomes here, if the magnets are aligned with the earth and kept at weak distance they can be used to tune the system altering its effective wire length on the tempic coils. 2nd outcome, Proton lines are severed and redirected, not completing there isotope lines at all and tempic resonance is lost for the wire entirely.

On Resonance

In a system using multiple coils in series there are theoretically two wave resonance parameters that need to be considered. The Earths Torsion wave is setting right under us interacting on 16 foot centers of its grid. It will resonate in a 45 foot wire, but this is torsion resonance, approx 10 Mhz, Proton frequency, totally undetectable to any EM sensing equipment. This wave can be canceled to produce higher density effects using the 1/4 wave canceling stub on a scalar coil. This requires folding the wire into four sections of equal length before winding it on the form. It can also be done using Sweets method, winding two coils in bifilar pattern each turning oppositely up the coil form. If the wave is instead converted back to EM by adding it's electric vector back into it, then we end up with an EM wave resonant at about 28 feet in copper wires or lengths of iron wire, or it can simply hit tuned circuits. A balanced tuned circuit would be ideal. The EM wave will resonate in any Electron active materials.

Chaining coils together will result in different resonance for each wave in the series length, but both waves must go through all the wire. The scalar coils are 45 feet long and sit at 45 or even 90 degrees to the electric coils. If one uses a combination of coils adding up to 316 feet, then a perfect resonance is achieved for both waves in the same length wire. Also there are ways to tune each wave that will not effect the other as much. Adding iron or steel wool to the normal coils will lower their EM resonant frequency but not effect the Copper length so probably not effect tempic wave resonance. Steel wool wrapped around the coils will make the EM wave path look longer.

A series capacitor will make the EM wave path look shorter, but since the plates do not touch [air gap] the torsion wave will not propagate without a magnetic field present to leap the gap. Tuning the tempic wave is pretty much based on the actual end to end Copper path. Adding a tempic tuning coil will extend the path, then taps can be added or a sliding tuning stub.

A fully resonant system will have to resonate both fields, and it is believed that this will then become very powerful, based on the amount of Copper present in the torsion field. Making coils bigger will increase the field as torsion flows through all the AG metals with no resistance. Core materials can be added also to increase the effects.

Here is a chart showing wire lengths for multiple waves of each kind.

Torsion Wave Length for Tera Waves

45.2548	1	
90.5096	2	
135.7644	3	
181.0192	4	
226.274	5	
271.5288	6	
316.7836	**7**	********
362.0384	8	
407.2932	9	

EM Wave Length for Tera Waves

28.81	1
57.62	2

86.43	3
115.24	4
144.05	5
172.86	6
201.67	7
230.48	8
259.29	9
288.10	10
316.91	**11** ****
345.72	12
374.53	13
403.34	14
432.15	15

As we scan down the numbers we see a very interesting crossing point for both waves at 316.7836 to 316.91 feet. A single wire this long will resonate both waves. There will be 7 standing torsion waves, and 11 standing EM waves along the single wire. This is the exact amount of wire appearing in Lyles 7 coils on the Star Link System. One could use 7 torsion coils wired in series or 11 EM coils and the same result would be present, however at the coil junctions would appear a different mix of the two waves. Also with coils the EM wave is lowered in frequency from capacitance depending on wire gauge used.

Shutdown and Control

Caution is in order when working with Earth Resonant coils. With density effects there is always a balance between two torsion systems. When one has built many Earth torsion resonant non scalar coils at 44 to 45 feet long placing them into a device, if the normal coils are left unterminated the energy at a stationary point becomes highly charged with alternate torsion forces at a higher density. The earth charges them all the same direction. Moving the unit will not immediately shut down the effect. Also anywhere you set them they will continue to operate creating a new vortex system. In an opened area this can cause headaches and tempic anomalies. To control this they must be reverse wired, or enclosed in Aluminum containment. The simplest method is to wire top of the top coil to the bottom of the lowest coil. The flows will now move through one another and balance both halves of the Neutrons torsion. The scalar coils do not have this problem because they are already wired two directions in balance.

It is also possible that to gain control and be able to shut these coils down may require opening the coils at a non resonant tempic point, and installing a switch in them. An off switch or shorting link may be practical. This is only necessary in the Earth resonant coils as they are pulling on the background Aether, but powered directly from the Earths field grid. If a system is built with 11 EM coils then simply opening them should break torsion resonance.

The basic rule is if you have any normal wound 44.5 foot coils of wire present, they must be dealt with to achieve a balance and appear in sets of two reverse wired, otherwise you will be creating a torsion force that may grow over time. If it hits a density threshold things may begin to disappear.

One can also add an extra length of coil to make the entire system non resonant, but simply opening the circuit will not stop the resonance of a torsion wave. The other alternative is to tear them apart one by one and cut the windings. This should be given careful consideration because a runaway density device can be troublesome.

Further Experiment
[2-17-2007]

Experiment with my unit pictured above shows that the coils wound with #14 wire have a high capacitance. EM resonance for the four normal coils was discovered to be at 1 Mhz rather then 10 Mhz as expected. The scalar coils act only as resistance to EM as well. So we see an art developing as to matching the actual waves that will be present in any one device and tuning equipment will be necessary probably for any system wishing to attain total resonance of both tempic and EM wave paths.

Martin has developed a method for tuning the Star Link system using a variable capacitor, a 100K potientometer, 3 switches, with wire run to the steel wool covering of the coils.

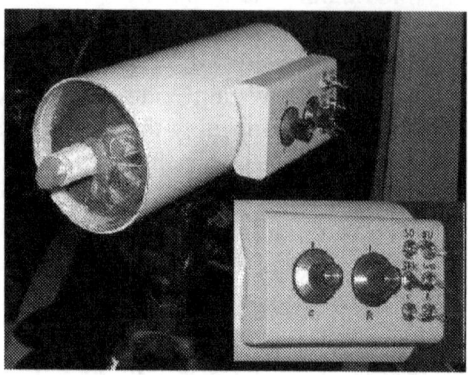

Martin writes:

By some tests with capacitor I found a real interesting position for it. As I first tested the capacitor inside the circuit of the coils, the output was weaker without it. Then, using the steel wool layer as ground shield, the capacitor could regulate the vibration between steel wool and coils. Therefore we need a little variable capacitor from an old pocket transistor radio and one 1,5mm thick copper wire, 2/3 as long as the big tube. Carefully stick the copper wire inside the steel wool layer from bottom bell of the device direction to the top, then connect this to one lead of the capacitor. The other end of the capacitor goes to the junction of rope and lower bell wire. Three lines are then connected there.

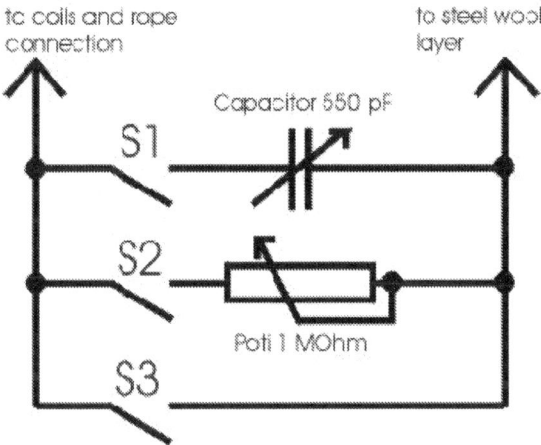

By this method we get 3 times more output! By turning the capacitor, the device radiates through various resonances or frequencies. I feel this moving up and down the spine while turning. If we shortcut steel wool with the junction, the device is nearly off. Absolutely astonishing this all...

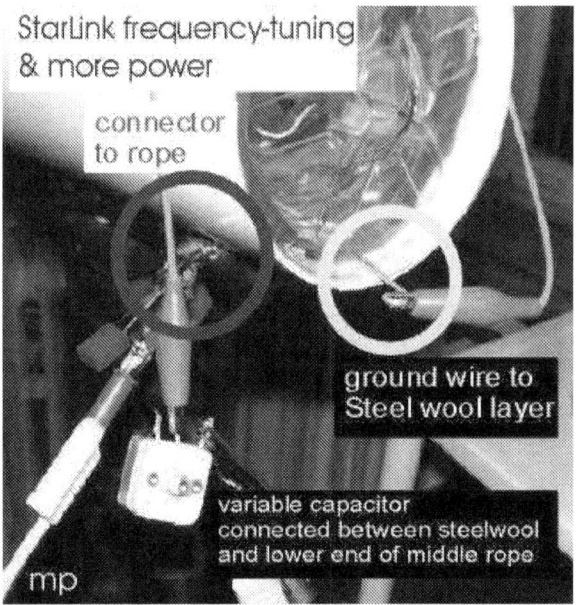

In addition to on frequency-tuning, it's the next point to control the output strength using an additional potentiometer with 100Kohm or more parallel to the capacitor.

If the potentiometer has 0 Ohm, the steel layer and the coils are shortcut, the device has minimal output. Turning the potentiometer direction to maximum resistor, the device powers up according to the level of the setting.

Additional 3 switches are good:
1) one for the capacitor to switch on
2) one for the potentiometer to connect to circuit
3) one for a short circuit between the steel wool with the inner circuit

By this method above we can control the device precisely.
Martin

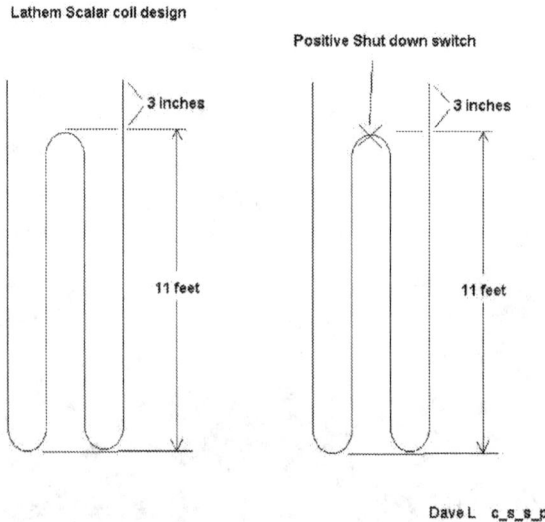

2-18-2007

I wound a Lathem scalar coil on a Bismuth slug core. The torsion is immediately present on the bismuth! Fanned out the extra looped end and cut it open, the torsion is felt to be altered immediately but over time reappears at a lower level. It was believed that placing a switch in the close loop end as shown in the diagram above, could be used for control on the torsion waves. When doing this do not add more wire to the coil but instead pull some of the last loop off the coil and add the switch. Both ends must be opened to alter torsion resonance because either one closed will cause a resonance to appear in the 44 foot length of wire, both closed sets up a density effect. A DPST switch can be added to the self powering coil to provide a totally earth powered scalar coil arrangement. It is now believed that to achieve positive shut down will require a different design to open the loops at a non resonant place in the coil.

Theory would suggest setting three of these coils in quadrature to effect a smooth density shift around a single point in space. As the earth grid has two intersecting along its surface, the presence of only one coil will not compress it along both dimensions and the headache effect may be present until crystal mass or other means are used to balance it. My next experiment may be winding three of these onto a sphere to see if a balanced density effect can be achieved.

Dave L

Credits

The Star Link system was designed by Marion Lyle Lathem, and shared publicly on the Crystal-devices group Feb of 2007.

The Crystal-devices group was organized and is operated by Sharfuddin, known to us as Sharf on that site, as well as the closed group c-d-p-g.

The Earth Grid system was studied at the Oregon Vortex by the owner of the site over many years who published a pamphlet on the grid system and believed the key to gravity was held within.

The applications found in this document were organized and cross referenced by David Lowrance of c_s_s_p group with support of many on the group as well.

Experiments conducted by Martin Pott of Germany in the first independent duplication we are aware of.

Torsion and Scalar Canceling Coil Basics for ZPE

While the focus of this paper is to preserve the scientific parameters of this new found energy form, one should remember that the Earths flow of life force is not to be regarded as anything less then sacred.

Caution

The ZPE coils tap the Earths grid systems, and this is a very powerful torsion field system. If one is not accustomed to working with torsion fields it can cause confusion, disorientation, and irritability if a field runs too strong too fast. The Lathem coils at 44.5 feet can be configured in the Kosol style density sphere. When grounded to the earth using a single wire, they produce a strong surge upwards in torsion and density. The sensation is the top of ones head feels like it is being pushed upwards. If any nonlinear device is placed in series with the ground wire like a light bulb, this can create instant pain, headaches, or confusion. If the scalar coils are kept in balance the sensation is very smooth and can be endured, but the desire is to identify, tap, detect, then down shift this energy to a more human friendly usable form.

We owe a debt of gratitude to Lyle Lathem for this amazing discovery of how to tap into the high frequency torsion force. However consider this is the field of the Earths grid to avoid sleeping on top of. It is commonly stated that it can become toxic for us if we suck in too much, and do not transmute it into a usable form. Every individual will have a different tolerance for this so must become aware. Self observation is not an option at this point, but a requirement. Conscious training is essential and emotional stability. The healing orbs are suggested if ones aura is not clear. Expect accidents and irritability to result until these devices are understood.

Sensing the Energy

Many devices have been covered on this site to date. The RainMaker vortex generator with a full calcite crystal sphere has proved to be the most effective for me. The scalar bismuth coil can be shorted then a single wire brought off the ends as a test probe for torsion in circuits and coils. They can be instantly felt from anywhere around the sphere when they hit the coil probe. Knowing what is going on inside a torsion coil is of the greatest importance, and if the energy is at a high enough level simply touching the copper with a finger can be very revealing as well. Torsion energy is sensed as pure Light at the third eye, but enters the body in many ways. Subtle levels can be sensed touching the sphere of the RainMaker. Close the eyes, begin to add descriptive words to the energy sensed and a picture will form very quickly of the energy type encountered.

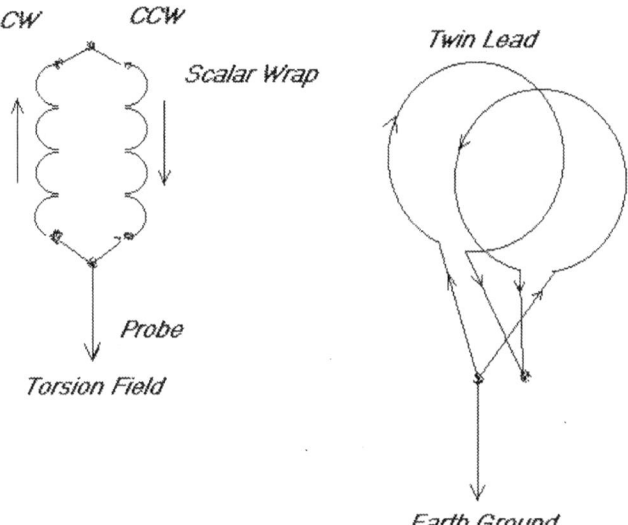

65

Also a large twin lead scalar coil can be used as a sensor of torsional energy, the ends of one side are run to a probe. Connecting the probe to a ground will demonstrate this effect nicely. Roll 33 feet of twin lead around a pop bottle, which gives a good human energy signature. Short opposite colored end of each lead together, solder a probe wire to one joint. As alternate torsion hits the coil it will be split into two flows that oppose one another and the energy will expand becoming perceptible at the third eye center. This is a one wire torsion circuit or opened ended circuit, and we see from this that it does transfer energy.

Earth Grids

Here is what is known to date on the earth grid system.

Curry Grid

The curry grid is of COSMIC ORIGIN - THE CURRY GRID, with a square of 3.6 x 3.6 meters. This grid has lines in 45 deg. to the main directions NWSO.The lines are 10cm thick, at full moon they rise to 30cm.

11.8 foot per side the grid sits at 45 degrees to poles of earth and appears to be an inflow to the earth. A copper shield held over a 44.5 foot coil will lower the coils output.

From this grid we derive the following terrestrial lengths:

Diagonal
11.8 feet
23.6 feet
35.4 feet
47.2 feet
North South East West
16.68
33.36
50.04

Hartmann Grid

The Hartmann grid is TERRESTRIAL ORIGIN - Hartmann GRID is the first grid, with direction directly NWSO, about 2.6m x 2m, with too 10cm thick.

Hartmann grid is also called global grid.

These lines form a North - South and East - West grid, they are about 2.0m apart on the North - South axis and 2.5m apart on the East -

West axis in the temperate zone. The active areas being lines 21cm wide with the neutral zone lying between the lines each 21 centimeters (9 inches) wide. The grid is magnetically orientated, from North to South they are encountered at intervals of 2 meters (6 feet 6 inches), while from East to West they are 2.5 meters (8 feet) apart. Between these geometric lines lies a neutral zone, an unperturbed micro-climate. This network penetrates everywhere, whether over open ground or through dwellings.

The Hartmann net has been defined using the Chinese terms of Yin and Yang. The Yin (North-South lines) is a cold energy which acts slowly, corresponds to winter, is related to cramps, humidity and all forms of rheumatism. The Yang (East-West lines) is a hot, dry rapidly acting energy. It is related to fire and is linked to inflammations.

The points formed by the intersection of these lines, whether positive or negative, are dynamic environments sensitive to the rhythms of the hours and the seasons.

It has been suggested that both the Curry grids and Hartmann Net are earthing grids for cosmic rays that constantly bombard the Earth, and that they can be distorted by other things, such as geological fault lines and underground mining. It is also possible to have spots where the Curry and Hartmann lines cross, causing further potential problems. These spots are generally seen to be more detrimental than a single crossing within the Curry or Hartmann system.

From this grid we derive the following terrestrial lengths:

North South
6.5 feet
13 feet
East West
8 feet
16 feet
32 feet
Diagonal
10 feet
20 feet

Note the 16 foot number popping up in both systems along the East West line only slightly off.

Testing

Our testing to date at c_s_s_p has correlated the 44.5 foot coils with the Curry Grid. The coils seem to pull in and collect this strong torsion energy. Moving them physically over the grid lines also shows this link as well as its origin, that lies somehow above the earth. The energy is Earth or ground seeking energy and will create a strong flow from the 44.5 foot coils if connected through a single grounding wire, probably 10 times stronger then without the wire.

Down Shifting and Using the Earths Energy Grids

Identify

If one doubts the reality of the Curry grid or its tremendous intensity, there is an easy proof. Wind a Scalar coil of 44.5 feet using two copper wires or a twin lead wire, then form the ends to join oppositely from each end of the coil. Now place a jumper from either side of this scalar coil to a good earth ground and wait. This is one that you will probably notice even if you do not consider yourself sensitive to energy. Destroy this coil and do not leave it laying around, it will pull in and concentrate the high energy or cosmic side of the Earth grid.

To make use of this torsion field we must discover a way to lower the frequency and split out the components we desire to use. This is something the Earth normally handles for us. This coil connected to RainMaker will turn the area toxic in short order. I have found personally an approximate two to four day tolerance for having one active. At this point I am highly emotionally motivated to cut it into very small pieces! I have done this with #12 and with #6 wire both on large 2" Plexiglass coil forms.

In the raw form this coil can be used to find the earth grid, and this will become apparent if you start to walk around with either a density sphere or a large one of these scalar coils while touching the bare copper or connect it into the RainMaker unit and hurry before the head blows off. Do not try to tap this energy system directly using diodes or light bulbs, any unbalance will cause high frequency emissions, that may lie up in the gamma regions. This is the closest thing to "radiant energy" I have encountered and must be respected.

Tap

Splitting the Earths inflow energy

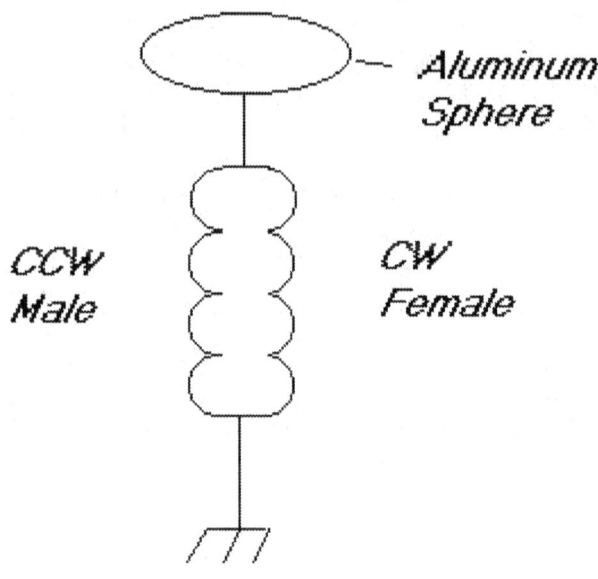

Aluminum Sphere

CCW Male *CW Female*

Cosmic Earth Grid

This is now the easy part of the system as Lyle has provided us the key. 44.5 foot coils of wire any gauge will isolate this grid systems energy and begin to suck it up. Multiple coils can be chained in series. Wind direction determines tempic field spin alteration. To increase the speed of this we ground one side of the coil to a good earth ground, even at a grid point. On the other end of this coil we may want to try a raised sphere of an AG metal like Copper Aluminum or Bismuth. Starting to look like a Tesla setup now. Spiral coils can be used as well. The energy is now taped using other coils that are not connected directly into the primary capturing system. They are positioned at the three spin planes of the system like in the density sphere. If one is made of iron it can be positioned with the earths magnetic field as well to shift the electric field to 90 degrees of this. All the forces will be split out of this, for two Cosmic energy sides, and wire lengths will determine the frequency on the output coils. It may be prudent to use only a down shift torsion coil and then route the energy away before pulling out the specific flows.

Detect

Adding iron anywhere around the coil will intensify the detection process. Sensors can be touched to the coil at any point, scalar coil or a density sphere, or just hands. A distinct difference is noted between the Female and Male flows. The energy is Earth seeking and touching it to a ground will intensify by 10 times in strength.

Down shift

The simplest form of a transformer would seem to be using shorter lengths of wire to tap off lower frequencies. As frequency is not setting in a coherent magnetic field, but a coherent torsional field, tempic energy will propagate between elements from Proton to Electron to Proton. Laying an iron layer over the 44.5 foot coil and a second shorter scalar coil over this may down shift the dangerously high frequencies to much lower ones. Now splitting the flows and then adding a torsion to EM conversion of 90 degree coils may make this energy usable as electricity. This is still experimental and our present goal.

While researching the Hubbard coil device one faces the first questions, coil size, wire lengths, coil spacing and geometry. These questions may be answerable from the scalar research provided here. To tap the Earths energy system down shifting is essential.

Torsion Versus EM Energy

Copper coil design must deal with two systems of energy. Normal EM is commonly understood and frequency drops as wires grow in length. As we wrap a wire into a coil the EM resonant frequency drops even more because of capacitance added between wires loops. With EM, the copper is acting along its Electron layer as a coherent magnetic field and thus propagation of EM follows the laws of antenna design. Shorter lengths raise in frequency and in Electron energy.

Torsion works entirely different and is based on the length of the isotope chains, and the mass of the copper. This is a coherent torsion field with non coherent magnetic fields in each atom. It involves mass of nucleus spin and precession vibrations. To make a 44.5 foot coil stronger with Earth torsion, use heavier wire, or double its length. Wrapping the wire into a coil does not seem to effect this at all, because the isotope chains in Copper appear to line up along its longer physical dimension. With shaping copper wire we are controlling the geometry of these isotope chains to effect its tension and side to side precessional torsion field interaction. The force is linear with distance and grounded already in the background Aethers time flow rate. Making the wire longer appears to increase the energy and probably the frequency also, this is opposite of the EM layer.

Clockwise Verses Counter Clockwise Torsion Coils

As we wind a coil into a CW wind we isolate the Female perceptive side of the cosmic force. Proton natural spin is opposed and torsion drops from the normal background level. As we wind the coil into a CCW form we isolate its male side. Proton spin is increased and torsion climbs higher then the background level.

So the female CW coils are seen to lower the higher energy state of the cosmic energy.

CW is viewed from either end of the coil moving into it, towards its' center. Either way energy flows through the coil it will be spinning the same way as to its direction of advancement, either CW or CCW. The two coils are thus unique and effect torsion differently to tap only one side of it. In a scalar coil we tap both sides and then connect them to cancel. With a torsion scalar coil the tempic field does not cancel but expands, there is no way known to cancel a torsion field at this point of knowledge. Tempic vectors in opposing directions always expand the field.

If we wind two of these each oppositely at 44.5 feet with #6 wire then connect the ends together, now connect one end of both to a ground, we will split the Earths cosmic inflow into two very strong energies. These coils lasted in this form for 4 days at which time I had to cut them up to tolerate the work space.

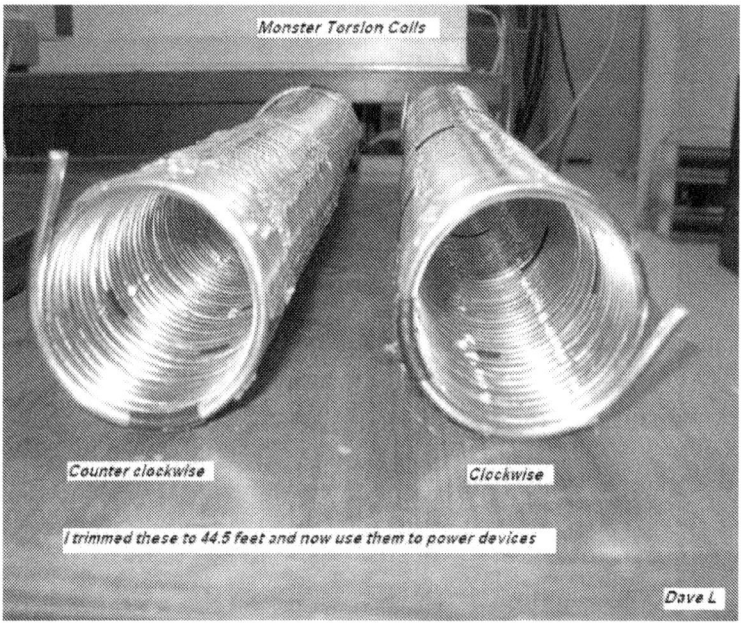

Monster Torsion Coils

Counter clockwise   Clockwise

I trimmed these to 44.5 feet and now use them to power devices

Dave L

So coils act like diodes to a torsion field in Copper. Turn direction determines energy type extracted. The two types of energy interact strongly.

Copper Torsion Conversion Factor

We see from Lyles 44.5 foot coils that the terrestrial distance of 47.2 feet is very probably the one we are tapping into. This gives the approximate Copper torsion ratio for design into any of the other systems. It is likely that a factor of .9428 be multiplied into the terrestrial length desired to draw in torsion field energy into the Copper. This places Copper wire lengths in the following chart for experiment. Trimming can also be used on site to determine this with a detection device like the RainMaker system.

Curry Grid

Diagonal
11.8 feet - 11.125 feet
23.6 feet - 22.25 feet
35.4 feet - 33.37 feet
47.2 feet - 44.50 feet

North South East West [as these may be magnetic they may follow the EM rules]
16.68
33.36
50.04

Hartman Grid

North South [as these may be magnetic they may follow the EM rules]
6.5 feet
13 feet

East West
8 feet
16 feet
32 feet

Diagonal
10 feet - 9.458 feet
20 feet - 18.86 feet

With the Hartman grid it is not yet known whether this has a torsional component strong enough to tap, however numbers are given for experimental purpose as coils that should be tested for Earth energy systems.

Combinations of 10 foot coils at 90 degrees in series have been stated to vanish or disappear on occasion.

David Lowrance - [Author and devices]
Martin Pott - [References and experimental devices]
Lyle Lathem - [Star link Grid coil system]
Kosol Ouch - [Density Sphere concept]
c_s_s_p experimenters group
c-d-p-g experimenter group

Field Density

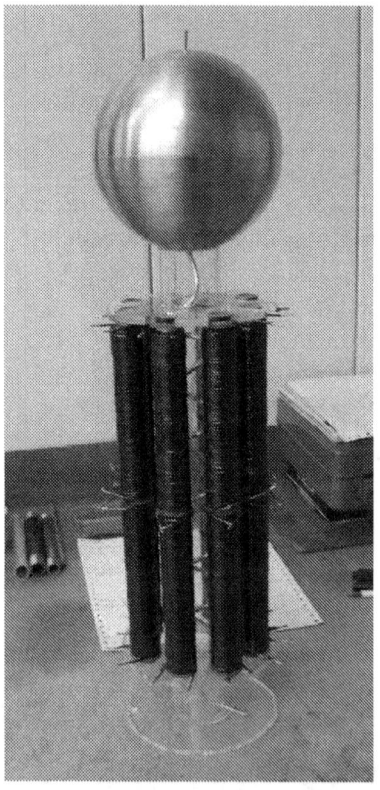

Shown above the experimental base for exploring field density in
Copper coils and Copper density spheres.

Tempic [Torsion related]
Electric [Voltage Potential related]
Magnetic [Current related]

Each one has a field density offering a different distance and motional range where the highest energy in its field will appear as a magnetic field is produced in Copper coils.

Tempic - linear fall off - 45 degree propagation angles - connects into all background time flow rates in surrounding matter
Electric - distance squared fall off - Spherical propagation
Magnetic - distance cubed fall off - spacial donut shape containment of major magnetic force present

Both the Electron shell and the Proton shell have all three fields in Copper atoms, however not in the same ratios. The Electron shell has a large Electric field, and the Proton shell has a large tempic field. The Proton shell is operating from inside the strong force area of the atom. In Copper only devices where no iron is present the magnetic field is greatly suppressed and we deal mostly with a Tempic Electric interaction. Without adding an iron core to a copper coil one would be hard pressed to attract an external mass of iron into motion.

Motional Ranges

The density sphere was created as a torsion device, to set up a higher tempic field density in all three planes of motion, using non powered scalar coils, creating a space of higher density at the center in balance. The density sphere, however, has allowed us to cross from the study of Torsion coils to EM coils and show the relationships for Magnetic, Electric and Torsional field densities as well.

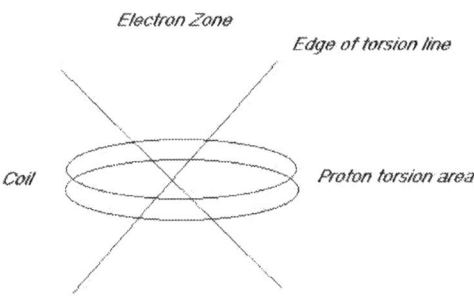

By placing an AC voltage on one normal coil of the density sphere, there are several observations of note. The result is an uneven spread of the component field densities and directions of travel. Direct experiment shows that in two coils at 90 degrees, one driven to resonance, in the second, a large voltage gradient will appear having almost no ability to produce current to a load. How can we make use of this?

Field Separation

As the Electron shell of the Copper follows the magnetic field into rotation, the Electric field and the Torsion field do not follow the same path across the sphere. The application of an AC wave to the horizontal coil of the density sphere will cause a magnetic field to begin flipping its poles inside the coil, and the torsion field will act on its electric component to spin it CW.

Electron action has almost 1800 times lower mass and will turn on a dime compared to Protons dominant torsion field.

As the Protons torsion will lag the Electrons faster rotation, due to increased mass, the fields split and spread their field density differently. Torsion pulls at 90 degrees in Copper and also counter rotates. At resonance in the coil both will cross in phase, but while the magnetic field is rising and lowering they will split outside the coil.

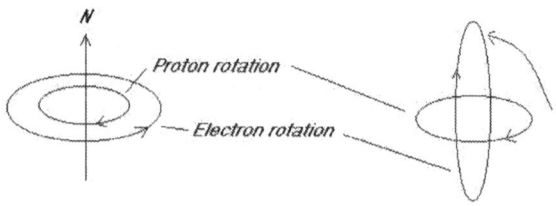

This diagram shows the counter rotation in Copper atoms between the Electric dominant force of the Electron shell, and the Torsionally dominant force of the Protons shell. At rest these counter rotating fields keep the atom balanced in a negative feedback loop. If we hit the coil with a fast electric impulse, the electron shell will tilt first, shown above to the right, and the Proton shell will lag this pulling off in a CW direction. The faster the rise time of the pulse the farther the two fields will separate. As they separate they will both experience a release of the negative influence of the other. Along the first plane of the coil torsion will surge into a reverse or back EMF pulse [causing a current flow back into the driving circuits of the coil.] In a positioned 90 degree coil, voltage will surge upwards free of the negative torsion force. Both will surge up free of the other but in opposing directions of spin. They are no longer interacting along the same plane of spin.

It is also good to note that at 90 degrees into the AC phase they will be pushing one another's spin planes, causing both to rotate by 90 degrees along the spin plane they would have been in if rotating freely of one another. Thus as both turn they both leave the original spin plane at different angles, but continue to interact with the fields that cross both as the angles change. At 90 degree of electric phase resonance, both have split into a different relationship with the input coil and reach 90 degrees, forming now 3 spin planes in space, all fully separated.

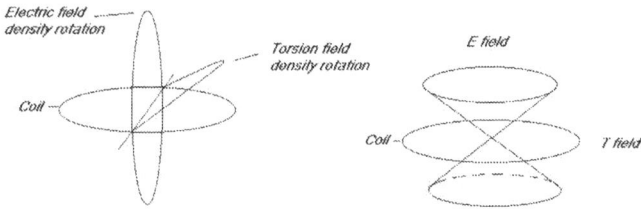

The torsion force will lag the turn of the electric field for an AC or pulsed copper coil. As the electrons electric field reaches the top of the magnetic field envelope, protons torsion will only have reached about half way up or 45 degrees. The electric field will now start back down, torsion will stop rising and start following the fall of the magnetic field back to the coil. The magnetic field will carry a component electric vector at the electron shell rotation as it rises and falls that will exceed the reach of back EMF.

We end up with Electric field density spread over the entire sphere, however at the upper 45 degrees it will be turning free of the reversed torsion fields of the Protons. The back EMF generating torsion force will remain along the 45 degree sides where a torsion line will form at 45 degrees, past which no torsion will be present. Torsion is a linear force and highly directional. This angle has previously been discovered in the tube devices where torsion is seen to maximize its flows along this angle and propagate with little loss if any through all the Copper mass present. This torsion line is also seen in permanent magnets as well and believed to represent the precession cones of the Protons motion, while now it may simply be the point along maximum field density for fields in motion.

So much for "eddy current" explanations being able to explain the torsional and reverse EMF action in Copper atoms. If the back EMF was a result of eddy currents at the electron shell, then the high voltage surge seen in the 90 degree coils would be countered by it, and not be present, as operating along the electron shell it would move with the incident field. The back EMF originates with the mass rotation of the torsion field of the lagging Proton shell, and here it can be separated and tapped independently [angularly] while in its expanded state of energy and its major field density separates from electrons reversed spin.

Identifying the parts of the magnetic field in motion that may offer over unity is now observed as the voltage surge zone along the upper area of the AC magnetic field where forwards EMF is released from back EMF, and the sides below 45 degrees where reversed torsion can be tapped directly and the countering EM field density is much lower. This reversed current producing force [Torsion] can be extracted from 45 degree coils and will also propagate Copper.

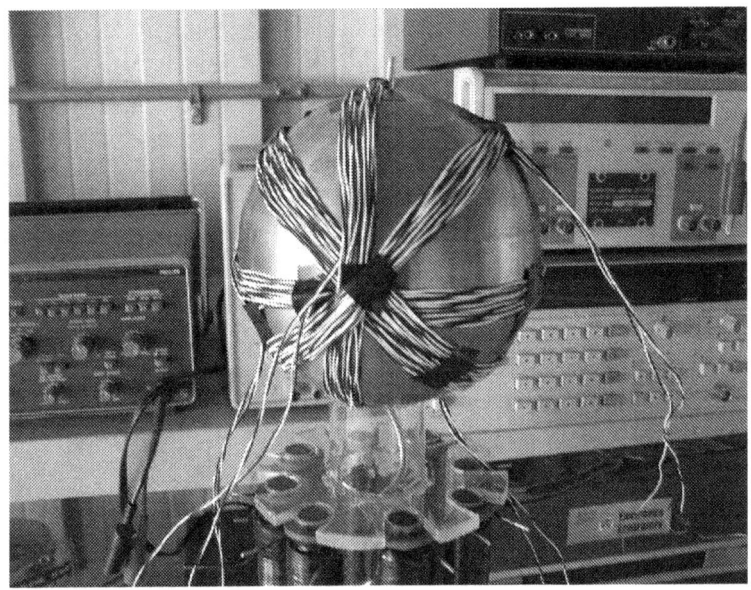

Scalar Coils

Now if we observe the interaction of two AC magnetic fields that are setting in a reversed or scalar canceling alignment, there is one major discovery made as their various field densities split and rotate through different planes of motion through the Copper density sphere.

A scalar canceling coil wrapped on the equator of a density sphere will throw off two voltage flows moving opposite directions while they each rotate CW. As one is reversed over the other, both rotating CW will cause them to separate and not continue to cancel perfectly. The torsion force will pull them both to a 90 degree position which will intersect in phase. This means that a scalar canceling coil will throw off a higher voltage vector then a single coil, and it will not cancel but at 90 degrees it will aid. Compounded to this the torsion fields will be doing similar but at a lower angle and begin to alter density along the lower 45 degrees of each side. As we increase the field strength of the AC wave we will see voltage nodes appear at top and bottom of the sphere. These are tapped in the scalar Capacitor of Otis Carrs diagrams, and the Copper and Aluminum split hemispheres of Kosols designs.

We can expect high voltage vector potentials to be available in a high powered scalar canceling arrangements, due only to the distribution of electric field density and torsion field density being separated and reordered in phase in the 3D of a spherical system. At the pole positions of a scalar canceling coil large Electric fields will be present. Along the sides of the scalar coils large tempic fields will be present. Both will be counter rotating to one another but in a motional state of separation.

The most obvious gain we find in this realization is that now, in non energized copper, where Protons all lie in a 50 percent magnetic alignment. A scalar coil will rotate the voltage vectors into an aiding alignment within 90 degrees of an input current applied. There is no magnetic field present, but a strong Electric potential will appear, as well as a strong torsion force.

This model shows why a scalar coil causes the magnetic field of the electrons to begin to move into a rotation around the equator of the sphere and diffuse into a blur of rotating poles in motion, while the diamagnetic field begins to become focused, along with an electric field mirroring gravity, having only an Electric and Tempic field actively focused. The gravity type field will appear at the pole positions and the tempic effects along the sides. Since the tempic field will be in a state of modulation as well as the electric field, this is not a pure gravity field, but a combination of both.

In a donut coil being driven by scalar coils along its outside ring, voltages will appear along the inner coil off the ends of each scalar coil, and torsion forces will appear along its sides. The explanation for this becomes evident from the interacting field forces present.

A Conical ESR - NMR battery

Microwave Resonator Area

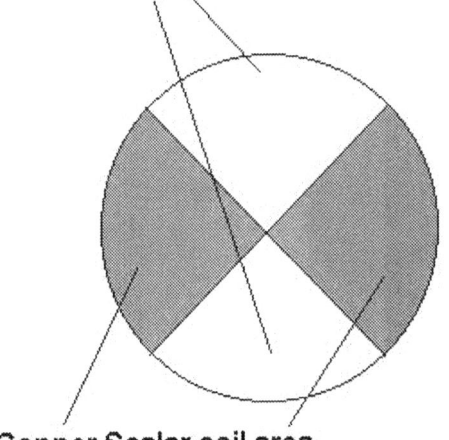

Copper Scalar coil area

Since the theoretical presentation of the NMR battery, several discoveries have been made as to field density and would suggest a better shape for an experimental unit to study the "over voltage" effect produced from tilting the Electrons spin momentum towards 90 degrees from Protons, and releasing higher energy into both.

The idea is to trap the Electrons spin into a tempic field alignment and resonate microwave energy vertically, while pulsing the Protons in the copper with square wave pulses that collide through one another. As the fields split closer to 90 degrees we should see progressively higher voltage present between the inside of two cones versus the outside sections.

Using cones with a "dielectric insulator" will separate these two voltage potentials and create a capacitor. If we hit on the correct pulsing technique then high voltages will emerge.

Negative potentials will appear at the top and bottom where no Proton E vector is present. Electron E vector potential will be released here to become very strong. Along the side of the coil, however far the Protons end up tilting, a positive E vector potential will appear.

A second conical copper or aluminum, or [alumina ceramic] insert can be placed above and below the hollow microwave resonator areas to act as a wave guide reflector to reflect a 1/4 wavelength and create a microwave scalar canceling layer. If this interaction creates a strong enough resonance, it will hold the electron shell rotation in place and set up standing waves in the tempic field. This will retard the electrons from recovering to their natural alignment countering Proton spin.

This sets up the equivalent of two scalar canceling systems at 90 degrees to one another, one at a microwave frequency and one at an NMR rate.

A simple version of this can be based on two scalar wound conical coils, with more scalar winding filling the area in orange above. A copper outer ring on the equator should follow the torsion of the system and produce the positive Proton or cold electric energy. The waveguide reflectors may collect the negative or electron hot electric energy.

This could be placed on two plastic funnels, large enough to provide the microwave lengths expected for the resonators. Previous devices would suggest a wavelength of around 2 to 4 inches, making a canceling reflector at 1/2 to 1 inch for a 1/4 wave cancellation of the fields. The canceled microwave fields will permeate all the copper present as a torsion force and synchronize all the copper atoms electron shells. The Proton torsion is already permeating all the Copper atoms present, so we should get a complete saturation of both control fields.

If a bismuth area is added inside the sides of the copper winding section a faster splitting is expected as the increased mass torsion of the bismuth will retard the Copper Protons field tilt. The bismuth will also offer a high resistance to Electron flows in this area. Since the microwave energy will be canceled it should be relatively safe when properly tuned up.

Windings on a Density Sphere

Many different configurations of coil windings can be experimented with on the copper sphere. Torsion will be coherent everywhere, and voltage gradients can be observed.

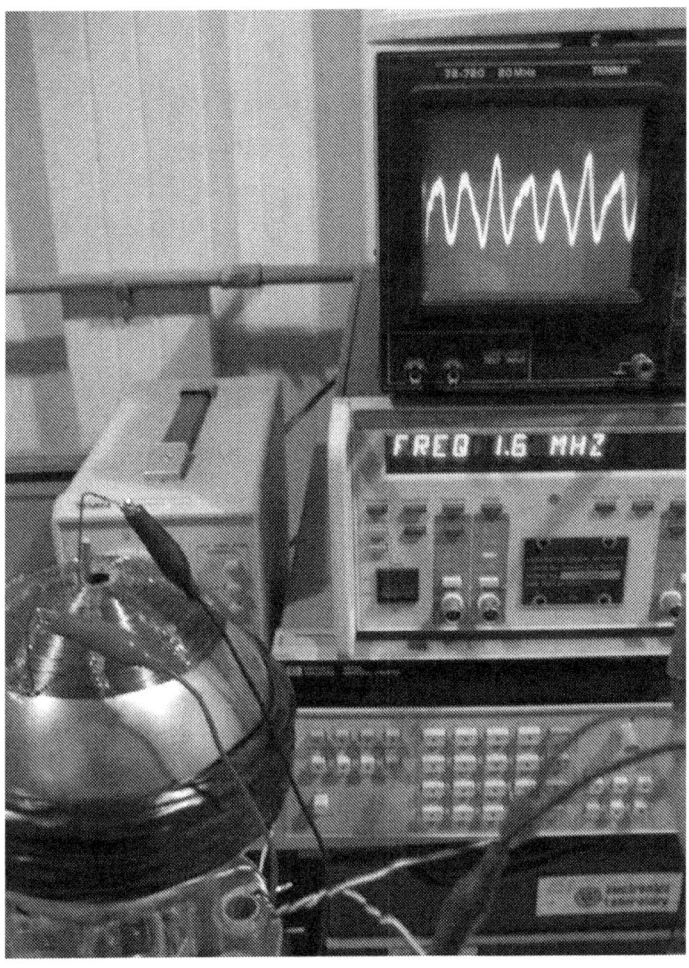

In this photo the large coil at the center of the sphere is being powered as a normal coil. The outputs from the top spiral coil are a single trace showing one wave is moving through the upper region carrying a voltage component shown on the scope, as it spins CW from the torsion vector during separation. Looking down from the top of the sphere, as the South pole flips up it will spin CW, as the North pole flips up is will spin CCW.

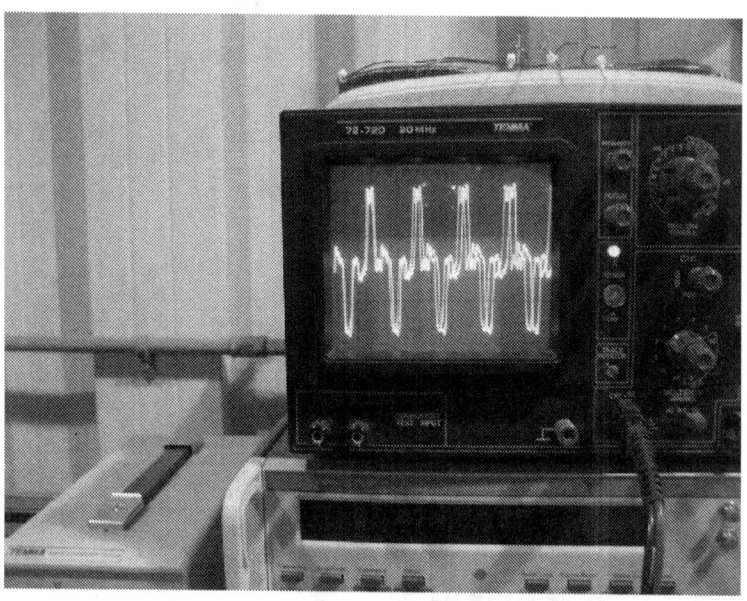

In this photo the lower coil is being driven as a scalar cancelling coil. Now there are two E vector waves passing over the top of the sphere, but the interesting observation is that, while they remain out of time sync slightly, they cross in phase and do not cancel along this area of the sphere. The voltage or E vector potentials do not cancel at the top of the sphere, as they turn and rotate they generate an over voltage component. This is tapping only one lead of the upper coil to show only its voltage gradient, ground is taken from the lower coil. The rise times for a scalar cancelling coil would also seem to be higher as there is a compression present along the incident coils angle. Fields are now pushing harder to get a way from one another.

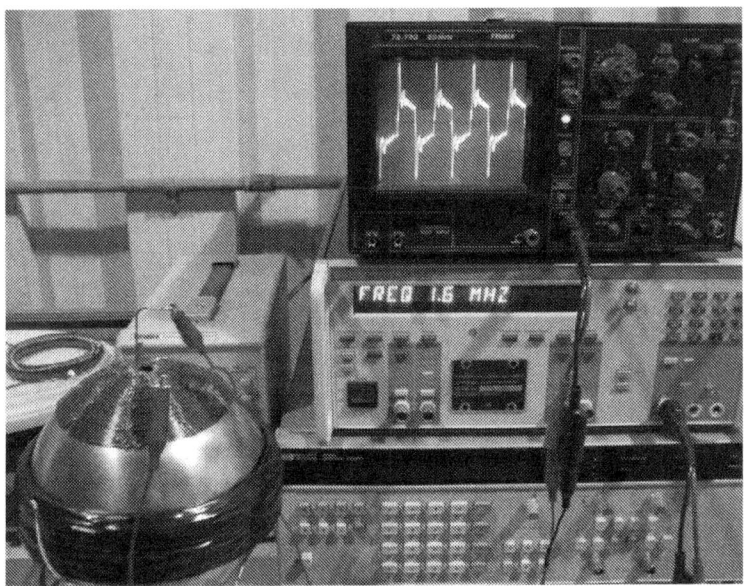

The scope trace shows an 80 volt peak spike in both directions for a 20 volt input pulse square wave. This is common when connecting scalar coils in various ways and hitting them with square waves. It is the Electron shell rotating into the upper area where Proton E vector potential does not counter it. This always leads the change of potential before the Protons drag can counter, and faster rise times will elevate this spike to unheard of voltages.

If we succeed at splitting the two fields over a longer cycle, then these high voltages we only see as short spikes will become present as voltage gradients. This gives one an idea how powerful copper atoms truly are. The main question is how much of this energy in these spikes is coming from the power supply and how much is coming from the Copper itself.

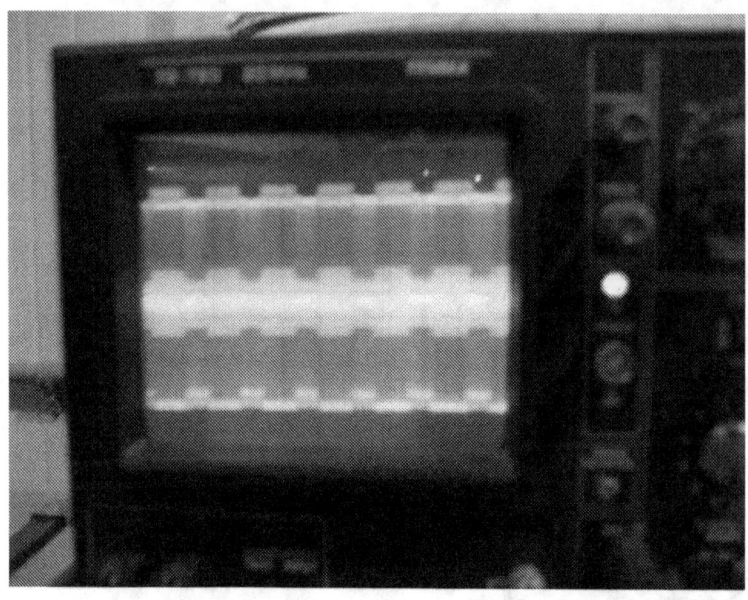

This photo shows the 5 K interactions that always appear when splitting the fields. Time base on the scope is slowed to show the Mhz waves interacting on the spheres surface where torsion forces seem to be pulling together then releasing at a regular interval. In no case do any frequencies have to add up to 5 Khz multiples, this seems to be a natural frequency from the copper itself near any resonance of the system for scalar coil arrangements.

Wave Forms
[Density sphere supplement]

Study of the complex E vector crossing top, with two square waves clashing near one frequency in the scalar coil.

Scalar coil on center of Density sphere. Driven by two function generators with reversed phasing near 1.618 Mhz. E vector potential tap at top and bottom of sphere is run into one set of coils along the bottom coil ring. First coil is magnetic canceling and E vector adding, second coil is E vector canceling and magnetic aiding, while both are Torsion splitting. This setup is capturing the voltage gradient at the top of the sphere and routing it to the lower coils that tend to isolate it by providing one side high impedance to a lower voltage driver coil stage.

The scope is displaying a complex pattern of the waveforms present as Copper Electron shells interact on the three coils. All waveforms originate on the sphere.

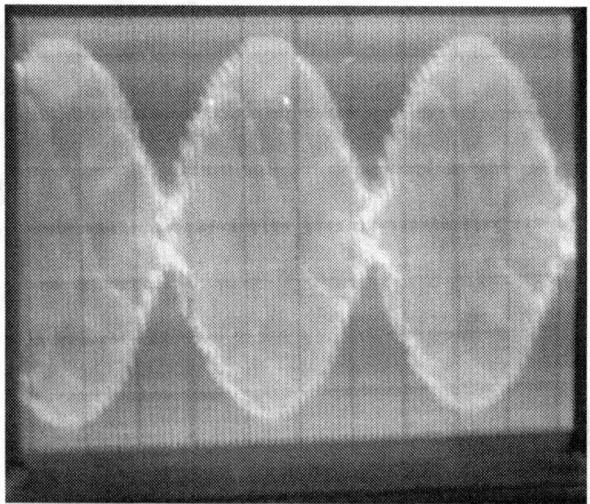

The composite waveform is around 1.618 Mhz moving two directions through the scalar coil. On the sphere these combine to produce several waves passing through one another, showing resultant lower frequency waves appearing. These resemble a phase modulation along the incident waves moving top to bottom of the screen.

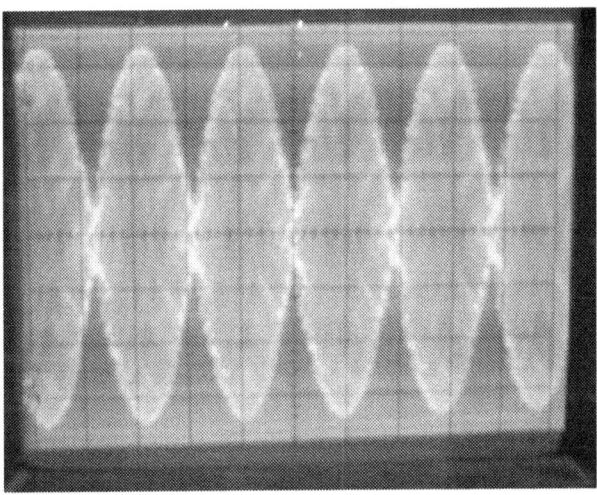

Scope time base lowered to show more pulses, wave forms are sine wave interactions.

Scope time base is raised to show the signal is really one trace with altered motion along the single wave, with the other waves passing through it.

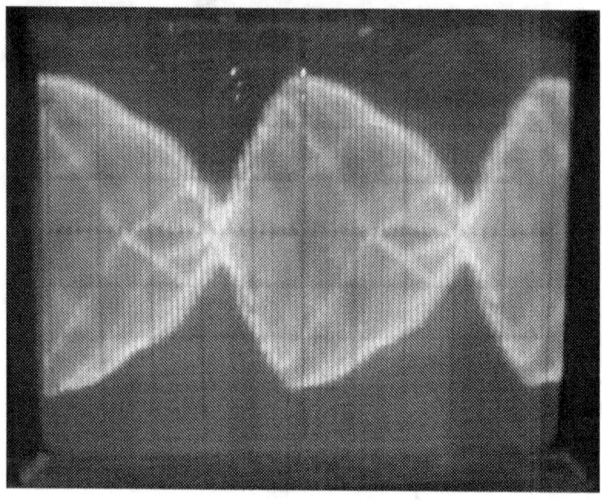

Here the lower coils are now being driven by the top spiral coil capacitor effect. The wave form begins to resemble the spiral coil itself in form. One lead is taken from the spiral coil and the other from the spheres surface along one side. This is all that is driving the lower coils.

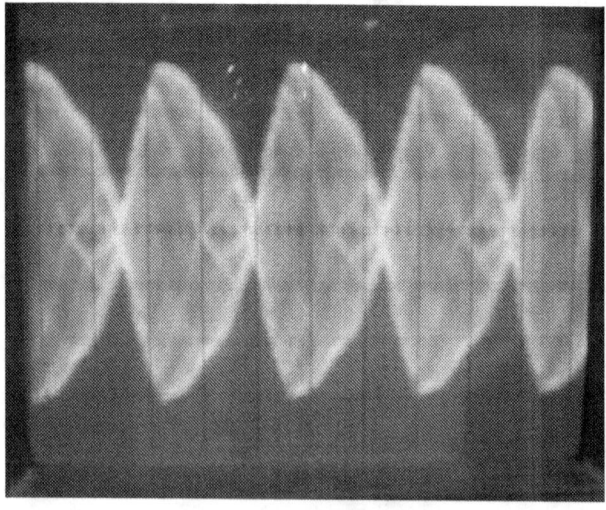

The top coil is adding a non linear wave shape to the spheres natural sine waves, and distorting the E vector potential to one side. This appears to raise the speed of E vector climb on the leading side of the waves growth, and retard its decline. We also see that the background peaking waves are eliminated and the output voltage has been amplitude modulated by this effect. Using two clashing driver coils we have extracted a lower frequency component from the sphere as a variable amplitude wave form. Pretty good for having no diodes or transistors, and no electronic mixer stages.

Scope here is 5 volts per division showing a 30 volt peak envelope. All this being transferred from the Sphere above from only a spiral coil at the top, to the spheres surface at the center. This is definitely a capacitive E vector potential effect. It clearly shows that within the operation of a scalar canceling coil, the voltage is not canceled off the ends but becomes an over voltage tapping point.

Over Unity in Copper

Preface

A device that appears to produce more output energy than goes into it can be loosely called an over unity device. However we know that energy does not come from no where. There is much speculation as to where this energy might come from, and many theories have been written now at the beginning of the 21st century, but to date no one is freely sharing a working device with the world. The purpose of this document is to show that this is very possible, and from what we already know today. One need only give up the notion that Copper atoms run on energy from the big bang winding down, and adopt a more rational concept that they are actually powered moment to moment from a light speed spin existing within them that produces continual mass momentum of the nucleus which is self regulating and self recovering back to a near light speed constant. A redirection of this internal force can set the stage to capture the internal energy.

In 1997 a device surfaced claimed to have been created by Steven Mark that was shown to produce some 1 Kw of power containing little more than Copper windings embedded in cork and plastic and some "secret" pulsing circuitry. The device was sealed, supposedly due to financial reasons. Once started up this device was self powering. One more example to be added to Floyd Sweet and Hubbard that appears to be over unity with respect to our physical world.

Theoretical

Looking closely at NMR [nuclear magnetic resonance] it becomes apparent that devices that pulse copper coils can be set up to produce more electric output than is put into them to cause the output that it should based on the simple turns ratio formulas, using 90 degree positioned coils. While this appears exciting on the surface we must realize there is a difference between the presence of voltage, and power, which is also a function of the current flowing. To cause an output from Copper wire which is driven, not entirely by the input pulse, but by the input pulse plus the atoms natural self regulating negative response to the input pulse and its normal mass rotation, seems to be the method. This can be though of as redirection of the atoms natural nuclear force. Wilbert Smith also spoke of reversed precession, found in Otis Carrs saga as forced precession. These are mass and motion manipulations, and when we compare the interaction of voltage and current in Copper we see the tempic field split the electric and magnetic forces into a leading and lagging situation. Normally this will not create an OU situation. The feeling of most studying the OU devices we have had access to, believe there is a condition possible that will in fact do this. The factor that is not being considered in the normal EM technology is the mass of rotation at the nucleus, and its natural precession motion. If this motion can be altered to lower the atoms negative responses than a tempic field acceleration may be possible. Copper has a magnetic hook at the nucleus allowing mass rotational effects to be manipulated. We have been finding this normally around 1 to 2 Mhz in many of the coil devices we are experimenting with. The actual frequency is a function of the magnetic field intensity, so with AC this is in a state of constant change and may follow the normal EM resonance of coils.

Density Sphere
[Square wave transmission]

It is not normally expected that a square wave can propagate a transformer because copper is seen conventionally to react only to a changing current.

In normal EM interactions, as we hit the electron shell with a flowing electric current, the Protons magnetic field will turn to align with it one of two ways, but it will always follow the magnetic poles bringing spin of mass along with it. The Proton will turn slower, however it carries the mass of the atom with it so the tempic field effect is a dragging of the current generated back out by the effect. This will cause the current to lag the voltage as the mass spins up, dragging the electron flow on the surface of the electron shell. The interaction resists the change of current and not its static state. Conventional EM electronics believes this is strictly an electron shell effect but will agree that the current will in fact lag the voltage in copper coils and this is a "real time" or tempic field effect. What they miss is that the effect is also a geometricly spherical effect and 90 degree coils can show this, they do not normally identify this with the coppers mass. In the 90 degree coil set ups we see extremely high voltage spikes as the voltage is turned 90 degrees faster than the Proton can turn the mass of the nucleus.

As we turn the magnetic field by 90 degrees, there is a reverse current surge along the original spin plane and a voltage surge along the new spin plane at 90 degrees. The two sit in spinning opposition to one another because one is the back EM and one is the forward EM. We have now separated the forwards Electron spin from the dragging Proton spin for a measure of real time at the NMR rate angularly. If we place two sets of coils on a density sphere at 90 degrees, pulse two at 90 degrees in series, then extract from two more in "series reversing" we get back out the original wave by capturing both spin planes and combining the current surge back into the voltage surge as aiding pulses rather than countering ones. The coil config will pass a square wave through it keeping the sides very sharp as it captures both surges. This is quite an accomplishment for a copper transformer. Passing a true square wave. We have captured both halves of this from both spin planes at 90 degrees and recombined them by reversing the phase. It is also possible to put an AC voltage into one scalar coil set and get the original wave back out by using the other two coils at 90 degrees to the primary. With these configurations we see actual power moving through a scalar cancelling coil and having enough current to carry a load. This is in direct violation to current teachings for EM as to cancelled magnetic fields.

Under normal circumstances the density sphere can give nearly a 1 to 1 power throughput on this and carry a load with a square wave moving through it. Very strange to see a scalar cancelling coil transmitting power to a load. Also very strange to see a square wave moving through a transformer, with flat tops to the wave forms, which are not thought to be a change of current, but when broken down at two angles in fact are.

In a system with one coil wrapped at 90 degrees over a larger inner coil to form a donut the following wave forms are observed. The system is pulsed with square waves of almost any frequency below 5 MHz.

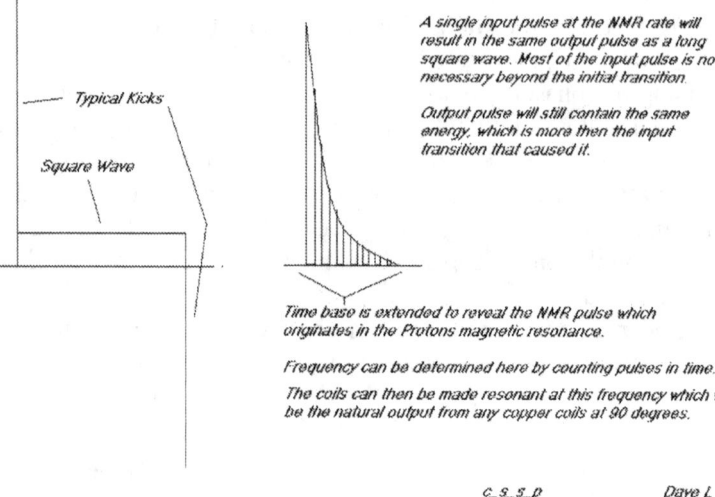

Pulses Observed in a two coil system where coils are at 90 degrees.

Shown on the left, the length of the square wave input pulse for the donut setup was altered to see what change it had on output and discovered that there only need be one pulse of around 5 MHz to produce the full output signature observed on the right above. NMR teaches that a 1/4 wave pulse is all that is necessary to set the mass of the copper atoms nucleus into motion. If one now adds these output pulses up coming back out of the 90 degree coil they discover it appears to have more energy over time than the single pulse starting the process. The output energy is directly related to the mass of the Copper under transition as it continues to move after the initial pulse is gone, and larger more massive coils produce more power output on this trailing ringing wave. Small coils produce unstable pulses and touching them can disturb these, but more massive coils produce strong pulses immune to external capacitive effects that become very stable. A single loop coil 6 inches wrapped with #10 wire, surrounded by 5 more #10 Copper rings produces strong pulses that are immune to touching.

The scope is showing the voltage surges, as the electron shell is released from the normal drag when the protons spin is aligned with it. The two forces are countering one another. We see the energy coming out along a different angular axis as the two shells are not in alignment.

The key to power extraction may be related to how the atoms of Copper recover, and it has been observed that the back EMF pulses are often more powerful than the electronics that generate them, often even burning out the driver transistors. While this process may seem rather random, it is not random but related to the NMR rates and when the pulses coincide. You may well pulse up a device many times and only on one out of 100 times the back EMF pulse will fry a transistor. This is probably due to NMR pulses stacking between the coils interactions. This is increased by adding copper to the core of the coils, or a combination of copper wrapped over iron. If done synchronously at the NMR rate I would expect we could learn to fry them every single time. We only need have two of the highest pulses hit at the same moment in time to double its amplitude.

Secondly it was amazingly discovered that even scalar wound coils still produce the NMR pulse just before the energy moves out of the EM layer. These kicks are ever present in scalar wound input coils as well as 90 degree coils wrapped around them.

It has been pointed out that using pulses at 90 degrees to one another does not create a true rotating magnetic field, however one must realize that at the Proton layer, the torsion is in fact in mass rotation. The rotation however is a spiral rotation and it is driven by the mass of the copper that is in rotation. While a pulse of EM through the Copper moves along its Electron shell, the spiraling rotation at the Proton shell is very real, and the tempic field is set now into a smooth sine wave spiraling motion containing a constant energy not totally related to the energy it took to start this redirection of its normal motion, but to its mass.

On the Steven Marks Device

Observing the rotation of a magnetic field through a Copper wire, where the field moves, not in the normal method of electric motors but between two coils at 90 degrees to one another in the donut pattern. Magnetic field rotates along the Electron shell, and tempic field swings along the Protons shell which is directly connected to its mass, a distance linear force. This process is an inertial momentum at the Proton layer but becomes EM as it hits the Electron shell once again coming outwards in Copper. This results in a magnetic rotation as well that can be measured with a compass at low frequencies.

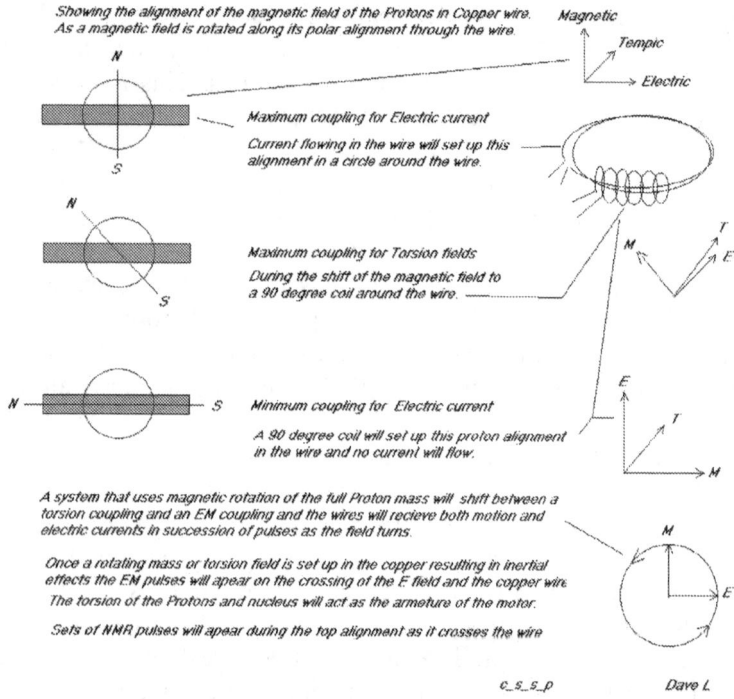

The one main advantage to using Copper is that it is the one element that will propagate both EM and Torsion fields through its length. The torsion flows most freely when setting at 45 degrees to the magnetic field, so as the field swings between each 90 degree angle there will be a spurt of torsion through the wire as it crosses 45 degrees. This will cause a physical vibration or a motional effect. As the magnetic field crosses 90 degrees to a wire we get the NMR pulse train an EM effect.

If it were possible to wind many coils at various angles to the input coil, one every few degrees, we would see an almost constant output of pulses in time as the field swings only 90 degrees back and forth between two pulsing coils.

360 Degree Rotation

If we can master rotation of the Copper nucleus in the correct spin direction, with the correct pulsing configuration, it may be possible to achieve a 360 degree spiral rotation which begins to counter natural precession, and constantly outputting more energy than it take to achieve this rotation. Observing lasigous patterns on an O-scope is an example of this in two dimensions. We must master a three dimensional one however to achieve a complete mass rotation and turn our Copper nucleus mass into a motor armature. If the recovery system of the atoms mass now moves into a closed repeating loop then energy can be extracted from the recovery energy that should far exceed energy to start the process. This is based on the observation that Copper atoms will always recover no matter the load we place on them, they always push back with stronger force.

An EM Density Sphere

A density sphere created on a solid or a hollow Copper sphere would be expected to attain a coherent torsion field through out, and thus using coils wrapped around it at various angles would provide start up energy as well as tapping energy. Output coils must be separate from input coils, and coils can all be driven with the natural outputs from other NMR resonant donut coils producing the correct frequencies of NMR pulses or very close to it.

In this special case of a density sphere, we are moving not only torsion fields but Electron fields into interaction. Voltage follows the Electron shell and current follows the Protons turns angularly. EM resonant coils can be placed at 45 degrees to torsion resonant coils if both interactions are desired to study.

Output is a direct function of Copper mass in rotation along the tempic field. Since the tempic field effects density and gravity as well, this spherical device will emulate a single copper atom along the torsion field, and a single coherent magnet along the Electron shell when Electrons are flowing.

This is the first motor suggested that contains a 3 dimensional armature composed of the mass of all the Copper present in the sphere. The sphere is a Copper only device.

It is believed that using a sphere will allow these principles of mass rotation to be accurately studied and documented. Three large and separate donut coils can be used to drive the system tied to three or more driver coils on the sphere at various angles. In this way independent or non coherent Copper driver coils can be used to pulse the sphere into mass rotation rather than an enormous system of pulsing circuits. Feedback can be via electric or tempic field depending on the coils positioning. 90 degree coils in series can rebuild tempic fields at 5 Mhz external to the device to provide feedback from a 45 degree coil, or 5 Mhz resonant EM coils can be used.

Control requires only three pulsing transistors that can vary frequency and phase. These are run into three separate donut coils and output will be NMR resonant pulses based on the mass of the donuts. To totally eliminate the possibility of frying the transistors the primary side of the donuts can be wound with scalar coils, this will eliminate any possible back EMF. The output side will be normal wound coils, 5 MHz resonant, and feed the next coil series.

Reversed Precession

If a density sphere is driven with DC pulses along the three planes of motion, the magnetic field will move into a precession motion. The North pole of the coils will move in a basic triangle without flipping over. Protons spinning mass will lag this turn. If the direction of rotation is reverse of the Protons mass spin this may lower inductive drag and accelerate the atoms energy state. It is very probably this sort of manipulation at the NMR rates that may lead to some interesting effects in Copper offering more energy out than goes in.

Density Sphere as a Transformer

With correct configuration of four coils on the surface of a density sphere we can create a transformer that will pass square waves as well as power a load.

With four coils in another configuration we can power a load using a scalar coil primary in only one spin plane and capturing voltage and current from the other two spin planes as it swings between them. This uses all three spin planes. As the scalar coil is powered with AC the voltage and current alternate between the 90 degree coils and can be recombined in the correct phases to add rather than counter.

This shows the mechanism of scalar cancelling coils and how the actual energy present is not lost, but moves into the inertial momentum of the particle spin motions, and then back out. It also shows how the elements of electronics are actually split between the two shells in operation. The electron being lighter and faster carrying the voltage aspect, and the Proton being slower but more massive carrying the current aspect of the force, in copper. The interaction between the two carrying the power, which is the product of both normally countering spin forces.

Dave L
c_s_s_p group
4 - 4 – 2007

Kosol Ouch, Koeun Noun Ouch, David Lowrance, Martin Pott,
Jerry Evans II and Vince Panella

Sweet DTO
[Diamagnetic Torsion Oscillator]

Below appears the legendary circuit diagram of the Sweet VTA [Vacuum Triode Amplifier] as labeled by Tom Bearden during his theoretical study of the unit.

Note the calculus, references to vacuum energy, and notes on scalar potentials. However one is immediately drawn to the fact that without current in the windings there could be no magnetic field. The concept of scalar potentials, or canceled EM is thus recognized now as a torsion field.

After study of Lakhovsky coils and there ability to remove the EM leaving only the torsion field which will freely propagate Copper. It was discovered that using two coils at 90 degrees to one another, wired in series, can be used to recombine the torsion from the Nucleus back into an EM wave, turning it back into electricity. This is apparently the function of FB1 and EX1. Thus the bifilar coils canceled EM energy may be recaptured from the torsion field.

The coil network setting at the center is flowing Proton torsion fields as they extend outwards from nucleus to the outside world. Placing coils at 90 degrees to a moving magnetic field isolates the tempic field, and in a scalar winding cancels all EM leaving only the component torsion.

The coils of the out side are flowing recombined EM as coils FB1 [Torsion] and EX1 [Electric] are recombining a small portion of the torsion energy in one side while FB2 and EX2 are combining part of the torsion energy on that side of the circuit. As each one recombines to form an electric field it then shoots to the other side of the circuit to power the opposite phase of the oscillator at the normal EM level.

As the EM field passes into the core windings it is canceled within a few micro seconds and the energy disappears from the electric field to become a torsion field. These torsion fields now interact between the scalar bifilar windings, which are setting in a specially treated magnetic field similar to the RainMaker base with a mono pole bubble at the center, and begin to interact as torsion fields at the proton layer of the copper atoms. As the magnetic field passes through these windings you will notice due to the wires alignment it will not capture the electric component but rather the tempic field at 90 degrees to it. The coils are setting at 90 degrees to the EM coils in the sides. This altered tempic field also propagates Copper as a hot or cold sensation, but disappears from EM equipment. The tempic field travels between atoms through the Protons magnetic field transferring torsion.

We now know that turns ratio is not the main factor in a torsion coil but copper wire length, and mass. The tempic field will resonate in Copper based on its propagation attributes, but its velocity is 471,240 Km/sec making its wavelength longer then in the EM coils. Thus 240 turns of #20 is almost useless information unless we know the length of a turn. We discovered the significance of 45 foot coils to resonate the earths torsion field. 4 connected in this manner will balance, but if left opened will create strong local torsion effects. Mass is important in the torsion effect also and heavier wire will deliver more energy. Torsion fields flow, without loss, to fill all the Copper present, but the energy of torsion is stored in the Copper itself. To have a stronger energy flow simply add more Copper of the correct length to resonate the torsion waves present. If the coils are all made to be 316 feet long each one will resonate both the earths tempic and electric fields. If they are some smaller resonate section around 44 to 45 feet long they will resonate only the earths tempic field and generate torsion directly from the Earth Grid.

Tempic
Electric
Magnetic

The three fields set in quadrature.

105

Since the tempic field is seen in many experiments as a torsion field, without its electric or magnetic component, it is seen to be the Primary force and has the greatest reach. The tube devices are examples of massive torsion field generators.

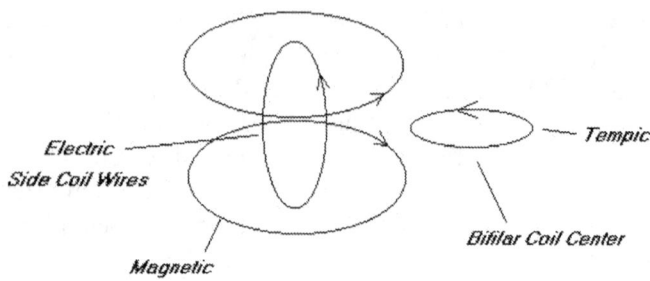

Torsion

The diagram above shows the flows of the 3 fields and the 90 degree spacial relationship of the tempic coils to the electric coils, exposing the nature of the torsion waves and how they are captured. The Bifilar coils in the center are not flowing electricity on the Electron layer, as bifilar coils remove this energy and push it into the Protons torsion field.

If one torsion coil is wound clockwise upwards while the other is wound counter clockwise upwards, I would expect an opposing torsion field to appear at top and bottom of each one setting together. This is from comparison to how the torsion field is received as a wave spinning along its edge, similar to a photon. Although this is guess work at this point, and whom ever drew this diagram was probably not aware of the possible difference. The torsion flows would move opposite directions in each one.

This being the case we notice the connection between them to realize that they are canceling the torsion field as well, and makes sense in this context. Torsion flows opposing directions through all the coils. This is putting pressure directly on the Aether, and expanding time, or causing a higher time flow rate. The torsion forces do not cancel but expand. The torsion field expands, although we may have no way to detect this with our

EM equipment and it will be invisible, with a linear distance reach. Either direction the torsion flows, as the magnetic field reverses in the outer winding, the torsion is opposing and so expands out to intercept the magnet fields.

As the magnets field above and below the device get hit with these strong torsion fields it is believed that it will alter their magnetic field connection. Since the magnetic field of a magnet is coherent, this will probably alter the precession motion of the field as well as field strength. We know that if the time flow rate increases enough then precession motions are lowered and gravity drops away. Whether the magnetic field from these magnets is beginning to swing strongly or not could be verified with EM equipment, however the torsion component of the magnetic field is definitely going to receive a pulse as it is passing top to bottom between the magnets, expanding and collapsing.

Oscillations

Whether the tempic field has any flexibility is not know, however in the torsion field of Protons and their spinning motions there may be a little, and torsion waves do exist. The oscillator may be oscillating at either side or a combination of both but the swings appears to be between the torsion field and the magnetic fields.

Torsion Detection

From study of the Sweet VTA as well as the Lakhovsky coils:

It is believed that a simple torsion to EM detector can be designed around two coils at 90 degrees to one another inside a Faraday cage. Coax coming out from the two coils wired in series can be hooked to many sensitive devices for indications. The coils will be directional and must be aligned with at least one coil parallel and in the torsion waves path. the second coil will reestablish the electric field and the magnetic field will result as the electric field begins to produce currents. So it is highly possible a torsion wave may be received and detected. This is the longitudinal wave spoken of on many alternate energy sites, but was never connected to Protons or torsion fields that I am aware.

Notes

Some of the interesting reading on Sweet by those trying to grasp his work turned up clips that can now be easily explained. A document by Walt Rosenthal states. The device went dead during an earthquake. No one could make sense of this other then possible EM fields produced. From the perception of torsion waves now being a reality operating in the tempic field, an earthquake would be releasing incredible ones. Other times it would simply go dead for no apparent reason, again torsion waves are not detectable except by sensitive people, and no one is publicly monitoring them.

The energy coming out did not loose power over distance and under load it never wavered although the coils had electric resistance. From 100 watts to 1000 watts the voltage did not alter. We have already observed the nature of diamagnetic fields in Copper. It is the Copper itself generating the fields as the Proton alignments move through it, rather then Electrons having to jump atoms. There is no physical transport of particles and no propagation loss, therefore no resistance in the wires. Energy is stored in the diamagnetic field of the Neutrons and transferred through the Proton isotope lines as motion, or torsion fields and then setting in every atom connected.

Probably the most interesting comment, that the energy can produce heat and power light bulbs that produce heat.

As the energy from the Copper moves into Electron based materials having no Proton magnetism or very little, it transfers into the Electron shell, to become Electric motion. At this point it becomes interactive with the outside world again closing the negative energy loop inside the atom. Possibly encountering any atoms that are not magnetic at the nucleus would produce this response. A study of the properties of tungsten may reveal the reason.

The unit must have a minimum load of 25 watts to keep working.

There is no flexibility in the torsion field as there is in the magnetic field. Placing an inductive link in the central circuit must be a requirement to maintain oscillations.

Dave L

c_s_s_p group

The New Electronics

This is an ongoing development document and experiment is underway presently to clearly identify the properties of the alternate coils and devices we have available. It is shared at this point to allow others to follow and learn. As information grows we will get there together.

Preface

Up to this point I have introduced the atoms self regulating energy loop and attempted to identify the fields of Torsion, and Magnetism. Showing the 90 degree operation between them, and how the loop passes through the Neutron operating at a higher light speed constant then the outside world. Now the goal is to develop a mental model for the standard Electronics Tech to gain a "feel" as to how to work with Torsion inside Copper coils and Aluminum barriers.

The Sweet VTA is a good sample of a basic model to consider. However new labels will be used. The DTO [Diamagnetic Torsion Oscillator] does not require we master any concepts of reversed time flow, or any other such complex imaginary 6 dimensional qualities of space time. It is based in the real world of 3 dimensions and the torsion component is setting in only one of these, a simple one dimensional force. Because the one dimensional force appears always as a circle or spin, the minds desire to over complicate this is amazing. No "Vacuum" suplying energy is required as well, we have identified the power source as the higher light speed constant operating within the diamagnetic field at the Neutron, of the Aether which is the same substance of the conscious planes, the Light.

The true nature of diamagnetism is not beyond a model of grasping the atoms reverse inner looping mechanism and identifying where the highest energy is truly setting. I will attempt to now offer Electronics vocabulary, a model for comprehension speaking strictly from what is already present in the craft. The field forces. Adding only one more concept, that of torsion.

Torsion

The tempic field force, a linear distance force.

Transferred through the magnetic field of the Proton when it has field connection with the Electron shell

Originates in the Neutron as the diamagnetic field having dual spin in a balanced state but always in motion

A dual torsion state is stored inside the Neutron and reflects the Zero Point energy state of the atom

Both Electron shell and Proton shell operate to alter this state, both in motion and interacting.

The torsion fields balance can be felt using hands or third eye center, but no scopes or meters have been found to detect it.

It has been referred to as the longitudinal wave, because this is how it propagates.

Electric

Transferred through the Electron in motion.

Set into motion by the diamagnetic fields torsion, if in the off balanced state, through the Protons and Neutrons interaction.

It propagates along the spin of a magnetic field moving in circles as in a magnet, and progressing along the torsion wave at 90 degrees to it.

It thus appears as a sine wave along a linear antenna at 90 degrees to the wave motion [tempic field].

Magnetic

Transferred through the Electron shells magnetic poles.

111

Tilted by the torsion of the diamagnetic field balance, transferred through the Proton shell.

Torsion in Copper

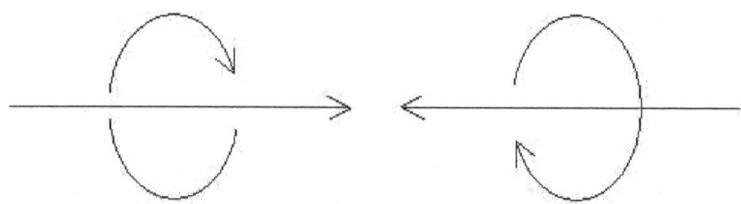

Torsion is the only concept we must add to Electronics to begin working with torsion oscillators. This is the part we have missed.

With respect to electric current in a wire what was missing is the teaching that motion is in fact a field force as well as electricity and magnetism.

The motion in electric motors however does not start in the wires electron layer but lies deeper in what is called the complex diamagnetic field of dual spin. Where there is setting two spinning forces in balance 90 degrees from the electrons directional current path as torsion forces. The outer shell of all Neutrons is the dual spin particle of the Electron. As we set either spin off balance then torsion starts to come up and move electrons at 90 degrees. Torsion is propagated through the Protons. This is the underlying force concept of Electronics [Electric fields] and Protonics [Torsion fields]. We do not have to understand the source of this interaction only to realize the reverse EMF or back EMF has a higher possible potential then the force causing it to come out, and it can be captured as a torsion field before it moves back into the magnetic field through the atom to close the negative energy loop. Also energy can be pushed into the torsion field where it will disappear from our test equipment until intentionally recovered.

The important concept to pick up here is that torsion lies just behind electricity normally spinning at about 90 degrees, but during opposing electric operations or static states, it hides. We were taught the current in a coil lags the voltage, and torsion is the reason. It is the time element of the electrons operations dragging the electrons and generating the inductive kickbacks on the collapse of the magnetic fields in any Copper present. The torsion release actually happens before the current starts to flow and it is the true source because it is the primary field force.

Torsion hides just before it strikes but in the DTO Torsion sets balance waiting to be released from a coherent state. The torsion is normally setting in a totally diffused state directionally in Copper so that when it is hiding, there is no way to detect its presence. When the Electrons start to flow then the torsion has focused into a direction of motion we can detect. Because it hides in a balanced state separately in every copper atom it is missed. So while in circuits we have capacitors and coils storing and releasing voltages and currents, in the Copper itself we have Torsion forces balancing and then shooting free setting electrons into motion at 90 degrees to physical motion of magnetic fields. While voltage is stored in the Capacitor, torsion is stored in the Copper itself. It is the off balance state of the complex Diamagnetic field that releases this Torsion. Capturing a torsion field requires special conditions be set up with copper. This prealignment in copper will cause the torsion to be released all in one direction creating the electric field. All our voltages push against this torsion force to tip it one way or the other way, the diamagnetic field is right there in Copper always ready to respond to this push and bounce back with spin, but while it is compressed inwards, it does not have coherent direction.

One can learn to work with a torsion force as though it were an electric field, the one difference, it is stronger because while electricity drops it's field say 16 times over a given distance, torsion drops it only 4 times as much over the same distance.

While the magnetic field is only present and coherent when we energize it, the Torsion field is always present and coherent through all the copper until we begin to manipulate it. If the Torsion field collapses in the Copper we loose its weight. This is an indication that we have pushed the torsion field to its limits and it is disconnecting with the outside world. Often Copper conductors will randomly be doing this but no one notices until they see a High Voltage line levitating during a random surge.

Opposing Coils

Placing a copper winding on a core material and then flowing current through the coil creates a push into the diamagnetic field of the copper atoms setting it off balance. As we vary the current in this coil now we get back a countering force of torsion causing an opposition to emerge, or counter EMF as it renters the magnetic field. This establishes the atomic loop in the Copper and brings up the complex diamagnetic field unbalancing its dual spin of torsion to one side. If the torsion reaches the electron layer then the negative self regulating loop is completed. We call this inductance in electronics, or Copper losses.

If we place two coils on the same core and now energize both in opposite directions, then we are now pushing the diamagnetic field in each one and storing torsion in both windings oppositely into two opposing diamagnetic fields. The magnetic fields cancel in this configuration and disappear, but the diamagnetic fields spin now splits between the coils and separates into two coils with opposing torsional spin pushing against one another on the tempic field. This sets diamagnetic field against diamagnetic field and the two coils have stored the torsional component field force in the Copper windings in a balanced state but now split physically. Although the magnetic field dissapears the torsion vectors are still pushing against one another and very much present. A one dimensional force.

If we use a nonlinear pulsing technique on the bifilar coils now we can isolate one side of this torsion and recapture it directly from the Torsion field rather then from the electric field. As the torsion field propagates effectively further then the electric field and its force is not countered inside the atoms until it interacts back into the magnetic field.

Separating the Tempic Field from the EM Field

The Lakhovsky or mobius coil shows us how to remove the torsion component from the Magnetic field. Using a pulsing method, one conductor flowing 5 volts and the second conductor pulsing 5 v to ground in the opposite direction, we see these coils can produce a torsion flow in copper tubes they are wrapped on. It can be felt with the hands and 3rd eye. A good sharp square wave is helpful in this process as it contains high frequency components, and the Proton layer operating mainly with torsion in Copper is not able to keep up with cancelling the effect, thus radiates the tempic or torsion wave.

Recombining the Diamagnetic and Torsion Fields to Rebuild a Magnetic Field

The Lakhovsky coils can be placed on a Copper tube, now two coils are placed at 90 degrees to one another, wired in series, and moved around until a perfect sine wave reappears from the Copper tube carrying only the torsion field generated from the mobius coil. This is much like what experimenters of the RF scalar coils report, and getting the X part of the coils to align perfectly allows reestablishing the magnetic field to the torsion wave. This technique uses two copper coils setting 90 degrees to one another to rebuild the electric field which now becomes visible on a scope. Comparing it to the square waves going into the mobius coil, they disappear right where the sine wave emerges. This shows us that the energy is moving through the copper tube as a Torsion field and not as a Magnetic field or an Electric field that we are used to measuring. It can not be detected with compass, volt meter, or scope in the copper tube or in the air anywhere around the mobius coils. Using the two coils at 90 degrees it can now be measured once again as electric energy. Now reexamination of the Sweet diagram.

115

The Torsion Wave

Longitudinal waves have long been recognized in the alternate
energy groups, there are many wild theories and observations
about these waves. Tesla spoke of high frequencies being
necessary to generate a faster then light wave. We have now
witnessed a wave that may travel without detection of any EM
equipment and then reassembled at the receiver coils. To generate
this wave we used a Lakhovsky coil, to reassemble it we used a
pair of coils wired in series, and setting at 90 degrees to one
another. One coil intercepted the Torsion or tempic field, the
other at 90 degrees reestablished the Electric field, and we ended
up then with the magnetic field, now detectable to our scopes and
meters.

Since the Torsion wave has a linear distance attenuation it
would be assumed this wave reaches much further then normal
EM waves as well it may travel faster. These properties remain to
be documented, but at least we have now devised a method to
study them, and layed a solid theory of what they are. Since only
3 dimensions are necessary the math should easily follow these
observations once the light speed constant is discovered.

When the higher light speed constant of the Torsion waves is
nailed down then a solid case can be made for Over Unity devices
using this method to produce power setting the two light speed
constants against one another, as is done in the Sweet VTA
[DTO]. Since our torsion waves are being emitted from the
Proton layer setting inside the strong force area of the nucleus it is
hopeful these will found to be traveling at a higher velocity.

Conductors

The best conductors of these torsion forces has already been
identified from the tube device experiments, as a copper
conductor setting inside an aluminum shell with a small air gap,
the diamagnetic torsion forces will then set up a field inside the
Aluminum containment and prevent it from escaping. This is the

new energy conduit for Torsion Electric Technology and lossless transmission. It is the Torsion conductor of the complex diamagnetic field.

Only diamagnetic torsion swing, off balance, will produce a cold energy. If perfectly balanced in the duty cycle, a perfectly balanced system will not produce cold but both cold and hot in a cycle. This energy will propagate the AG metals. Since the Sweet VTA has only one static magnet programming with one pole turned inwards and one outwards it could not alter this, and thus created only the cold half of the torsion fields balanced center. What we in the RainMaker field call the cold energy side or the "inflow". A properly balanced system will also have a cycle of "outflow" and it should still produce over unity power.

DTO
[Creating a Diamagnetic Torsion Oscillator]

Step One
[Creating an electric to torsion transfer without back EMF.]

As we will need a medium for transferring torsion we have selected a copper tube about 1 foot long.

At one end of this tube we place a Lakhovsky coil with two tightly twisted conductors, forming these into two separate circuits.

This is the input coil and converts electric pulsing into the torsion field of the Copper tube. Pulsing methods were covered in a previous document.

If set up correctly it should produce either a cold energy "inflow" in the whole length of the copper tube, or a hot energy "outflow" and be able to toggle between these on each cycle of the wave pattern applied to the two windings.

The reason this coil is used is that it will now offer total immunity to back EMF and not fry the power feeds or function generators when the large back EMF pulses appear in the system.

This coil will literally make the EM disappear going into it. The energy is transferred directly into the Coppers diamagnetic field and alters the torsion state of all the matter in the area, however it freely flows through Copper and Aluminum. The two together make the ideal conduit. An Aluminum tube is placed over the center area of the Copper tube between the coils with a very thin insulator on each end, this could be a layer of electrical tape at each end.

Step Two
[Creating a torsion to electric transfer - harnessing the Back EMF pulse.]

It is expected that now two kinds of energy may be tapped directly from the diamagnetic field once set into a state of pulsing in a tube of Copper. A Proton generated energy flow and an Electron generated energy flow. Sweet may have shown us each in different parts of his VTA. This energy stored in the diamagnetic field will move into any coils placed around the copper tube and can be studied.

If it is desired to transfer the diamagnetic field between two isotope chains that will form in the copper tube then a larger tube can be placed over the first on the output end and an Aluminum linking tube overlapping both at the center joint. Insulating tape is placed so none can touch. Now we are free to realign the isotope lines in each tube to different directions yet the diamagnetic field will pass between them. The output tube can now have a proton steering ring setting magnets at 45 degrees, as in the RainMaker base to see if a recovery system is possible. The isotope chains in the second tube will move into a 45 degree alignment and transfer energy between the tubes. Now a normal coil may be able to extract it as electric energy in a coil around this tube.

If a Lakhovsky coil is used as an output coil all we can expect to come out of this will be the Proton flows as it will always operate to remove all electron actions cancelling them. Two Lakhovsky coils setting together or interwound. One acting as the input and one the output. A bias current applied to one circuit of each coil then pulsing applied to one and output from the other to a load that can detect proton currents [light bulb]. The bias current is then altered in each to see if OU may appear.

Oscillator Function

It is not possible to oscillate the torsion field that I am aware. Torsion is a non flexible force and it expands the tempic field altering time flow rate, so oscillation must probably happen at the magnetic level in the electric coils and possibly capacitors, inductors or even magnets magnetic fields. As a torsion wave may alter the output of a permanent magnet, this possibility should be considered.

The idea now is to gather enough energy from these output coils to feed another identical tube passing back, but at enough distance so the magnets do not get near the Lakhovsky coils. Now we power this second one from the back EMF of the first one in such a way to produce the opposite flow. The back EMF spikes should be very sharp pulses and very strong. If tube one is producing an inflow pulse then tube two will produce an outflow. The two torsion fields will now be swinging back and forth between the tubes.

It is assumed that on the two output coils they may end up some 90 degrees out of phase. This is because the input pulse leads the back EMF pulse by this amount, but it should at least be one pulse width behind it. Input pulse on - one coil fires - input pulse drops - back EMF coil fires. These pulses need to be sharp square waves on the input. The back EMF pulse will be sharp.

The feedback loop at the electron layer may be enough to establish oscillations and light up a bulb if OU is found to be present in this system as is expected.

Over Unity

Over unity in the physical plane is not creating energy from no where, but tapping the higher light speed forces present at the nucleus of the atoms from the diamagnetic field. This force is

119

manifest a from the Proton layer as Torsion and permeates all matter. Whether one embraces the Density model of the conscious planes or not, we have already observed a lowered mass of the nuclear particles together, then if separated into their component parts. The photons of our visible light spectrum originate within the operation of the Electron shell and are subject to a lower light speed constant. Experiment will reveal if our line of progression will be able to harness the higher energy state of the Protons interactions.

2 - 6 - 2007
David Lowrance
c_s_s_p group

Precession and the Isotope Line

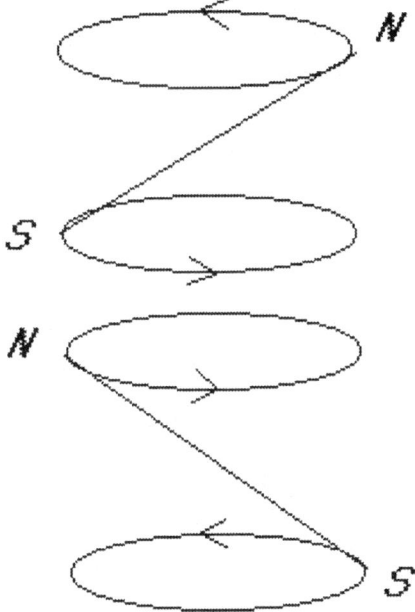

From the study of NMR technology one gets a feel for the nucleus of the atom existing as a free floating magnet, spinning its' ends like it was a bar magnet moving through a dual cone system. This motion is known as precession. With the AG metals Copper, Aluminum, and Bismuth the Electron shell is neutral and the nucleus of the atom is free to be manipulated using external magnetic fields.

As we bring a strong magnet, say a 1 Tesla Neo near the edge of the AG metal the nucleus's in the copper all align with the field. You can think of them as wanting to link together on the ends and move in opposite angles of about 45 degrees as they precess, chained together only at the ends. The small magnets are not strong or close enough to create a coherent magnetic field and each one continues to have separate poles and a separate blotch wall. They cannot become coherent with the external magnetic field from the Electron generated magnet because the spin is opposite with respect to the poles. These little magnets continue to remain separate bundles of energy self contained, and transferring energy independently. This is called the isotope line.

Cold Energy Configuration

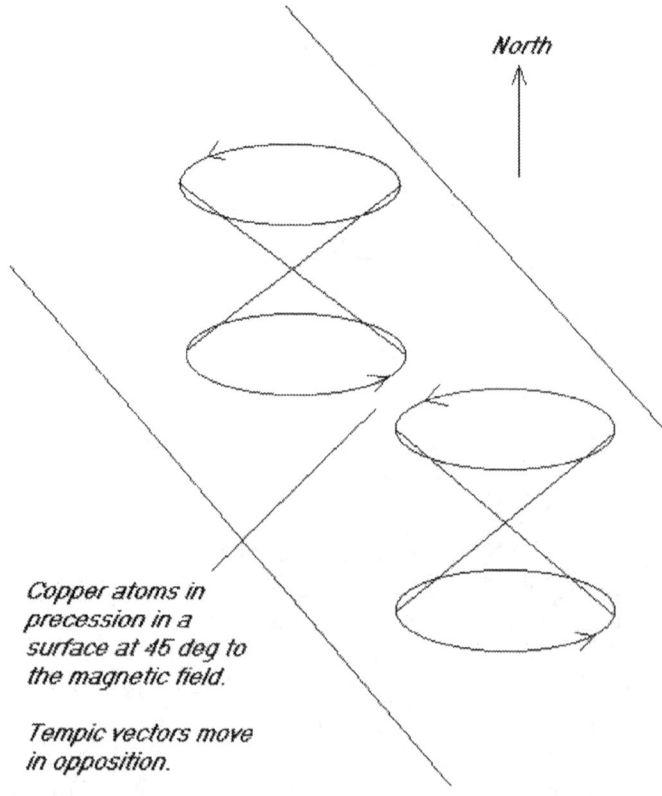

North

Copper atoms in
precession in a
surface at 45 deg to
the magnetic field.

Tempic vectors move
in opposition.

As we now start to tilt a magnetic field along a smooth thin surface of Copper or Aluminum atoms, the magnetic field manipulates their small independent precession angles such that they align with it. A greater number of the atoms now find themselves in the above alignment then in the first one shown.

In this alignment you will notice that they now spin past one another on opposite edges of their magnetic motions such that their motional vectors oppose.

This tempic vector alignment is recognized from the time vector model of magnetism and should tend to produce a higher density or expanding field in at least one plane of motion speeding the time flow rate.

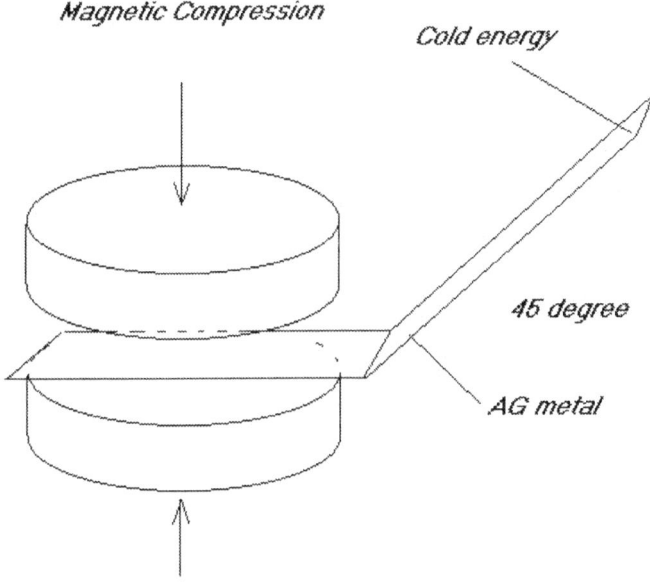

In the above model given, opposing magnets are added to the model to present two external magnetic fields also in opposition to one another.

With this system added to our AG metal, as the magnets are compressed both fields interact. Copper atoms can align two directions to one another and as the energy state of the nucleus moves higher more of the coppers nuclear magnetic fields will also oppose, however they will still precess against one another.

We end up with all three tempic vectors in opposition in the AG metal that is shooting off and up at 45 degrees in a high number of the atoms present.

A cold energy collects near the end of the extension piece which is colder further from the magnets.

Note that it is not the interacting large magnets in opposition that generate the tempic and diamagnetic effects. It is only after they interact in the AG metal that we see these effects popping up.

Experiments at c_s_s_p group showed this model to be fairly accurate, as a device using large ferrite ring magnets demonstrated.

We have already seen how once an alternate density field is created it can propagate seemingly without loss through the AG metals.

Diamagnetic Field Interaction

The diamagnetic field sits normally at about 90 degrees to the Protons magnetic field, however propagation is through the 45 degree vector of the precession line in the AG metals. Emitting from the Neutron it is also spinning somewhere between 90 degrees to the Protons spin plane and some angular balance point stabilizing all the atomic shells. It's angle is not fixed but regulates the atoms equilibrium.

Any time we set Protons against one another with respect to torsion we alter the tempic field and effect the time flow rate.

The ~ 45 degree isotope line creates two observed interactions that freely propagate. The first has been mentioned already the tempic effects which spread out at greater distance. The second is the diamagnetic field, which because of the torsion forces, pops out of the material and propagates along with the tempic field in an expanded mode down the AG metal. This is a conscious field mostly, although at its source it is a raw force that repels both ends of the magnetic forces present inside the atoms.

We see the conscious field coming up in too many devices now for it to be ignored. Hamels cones, Otis Carrs saucer, and a host of others as well. It would appear to be a necessary element. We are seeing many other reports now as well of the cold energy popping up as well as the conscious link in the RainMaker 1 unit as others come on line with new units and continue to experiment.

Neutron and Density

The heart of the diamagnetic field, the pathway to a conscious link, is the neutral center of our local system, the neutron. While it appears to be neutral to us in science, it is a very long distance from the center or Source of the universe in magnitude. With its center point lies the light speed variable, not a constant, that creates a possibility of density shifting. The nucleus of the atom is lighter then the sum of its parts removed, which can be easily verified by studying an atomic chart of the elements. The nucleus of the atom is already effecting a gravity shift from this perspective and thus creates a second Aether realm where light speed was measured already by Tesla to be higher then our normal Electron oriented space. Energies moving between nuclear bodies may bypass the normal EM routes taken by electronics generated vibrations as the medium is different.

David Lowrance
1 - 16 - 07 Kosol_Core_Tech group

125

Kosol Ouch, Koeun Noun Ouch, David Lowrance, Martin Pott,
Jerry Evans II and Vince Panella

Torsion Bonding

A closer examination of atoms and the Spinning Cylinder Experiment will show us all the paths of torsion outwards in the atom. As torsion couples from a Proton magnetic field to an Electron magnetic field, in the experiments with spinning cylinders, it does not couple between Electron shells in iron or between two copper cylinders in the same magnetic field. The atom must work the same.

In an atom all the normal electrons are bonded to the protons by a magnetic link and they are all paired. In the free Protons of copper we have one free to interact with us and show us how this works. As well we have a free Electron in the Electron shell in iron atoms to experiment with.

It should become apparent that all paired Electron Proton bonds also flow Torsion, although all these fields pretty much cancel near the surface, so how does torsion really flow between atoms. Torsion moves outwards in all the atoms magnetic links hopping from Proton to Electron to Proton to Electron etc... To get between atoms it must use the Electron shell as well as the Proton shell. This brings us to the molecule and the chemical links.

I has been observed that when we break up elements to the very small sized they often lift into the air as though weightless and are carried by the wind, even lead atoms and copper atoms if broken down small enough will vanish in the wind. Water seems to violate the laws of gravity in clouds as well.

We end up with a massive weaving between matter in all directions for all fields that appear at the atoms shells in the materials that are bonded at the Electron layer into compounds, we end up with a permanent link for torsion to flow between atoms. The Torsion field expands and begins to connect with

126

outside influences. Proton to shared Electron to Proton. This obstacle would seem almost impossible to overcome without breaking the bonds. The crystalline structures offer a possible link because they are all bonded along common lines, and thus pure crystals or crystalline metals will offer a structure to make the forces fall into atomic coherence if correctly aligned. The correct platonic forms lined up with the crystal bonds may offer a solution by introducing reversed torsion fields at the correct alignments.

The diamagnetic field is setting just below the Electron shell and it is holding all the Electrons in their orbits such that none are pulled into the nucleus. We have identified this with the neutron and its force function in matter to repel both ends of any magnetic field outwards. Proton is bonded to Neutron torsionally, as time flow rate is different at the neutron it is pushed away tempically, and the diamagnetic field pushes both Electrons and Protons outwards, while the Protons pull the Electrons inwards through the electric field. Diamagnetic field is pushing on both shells so it is present. A fine balance!

From the above logical flow it may become apparent that to decouple the torsion bonds in matter would be far easier if we had atoms that were not chemically bonded into solid matter, say in a diamagnetic liquid. The key lies in the diamagnetic field and setting up conditions to increase its reach such that it cuts off the flow between atoms. To do this it must expand outwards of the Electrons shell enough to reduce the torsion bonding between the atoms along its entire surface by canceling the magnetic field present there to such a level that torsion disconnects between molecules. This takes us to the simplest form of anti gravity device, a diamagnetic liquid based core.

While in a Copper device the diamagnetic field must interrupt the links constantly as they will immediately recouple over time, it must also be much stronger then if there were no chemical bonds. Copper atoms form into rhombic structures sharing a few Electrons with other bonding metals and the atoms or molecules are not free to spin at the Electron layer.

The simplest molecule and most abundant is water. It is already violating gravity effects in the clouds. Now with a greater

127

awareness of Torsion bonding and a better feel for how to control things from the conscious side, we are ready to discover the true method. Once a fluid diamagnetic material is released from the local Torsion coupled field it can now be manipulated to shift density using reversed spin coupling within a self contained torsion field and a conscious connection. The alternate torsion should propagate Aluminum or Copper due to previous experiments to spread this field into a craft uniformly.

Torsion Bonding in the Neutron

We look closely at the Neutron, it is a Proton with an Electron overlapped. As we interact it creates fields that regulate the atoms structure.

Since the Proton sets inside the Electron, we can now see how torsion is propagated inside the neutron. Proton - Electron - Proton moving out.

Torsion cannot move directly from the Neutron to the Electrons, only a reverse like magnetic field can, it must pass through the Proton layer. However this means that a Proton field interacting with a diamagnetic field will provide Torsion, but an Electron field interacting with the diamagnetic field will not.

This is key to why releasing the torsion coupling between atoms will not destroy the atoms. Torsion is always present at the Proton layer and its electric field will hold the Electron shell on as electric field is a distance squared force. Even as the magnetic field is overpowered by the diamagnetic field the atoms will remain together.

This also indicates that operating in a diamagnetic field environment, Copper becomes the new element that is interactive and magnets become only a repelling force.

Atoms and Over Unity

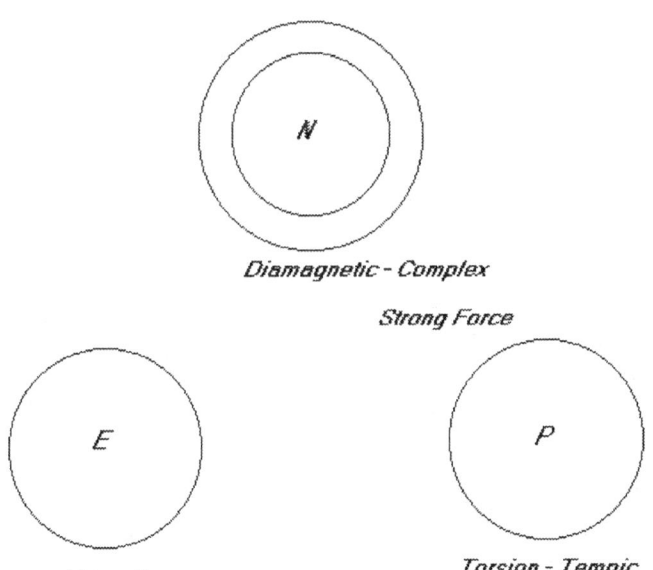

Diamagnetic - Complex

Strong Force

Magnetic

Torsion - Tempic

Review

Copper - Proton Magnet - Non coherent Magnetic - Coherent Torsion fields.
Iron - Electron Magnet - Coherent Magnetic fields.

Electron Proton –

Orbital attraction - Particle Repulsion.
Orbital repulsion - Particle Attraction.

Proton becomes coherent particle to its own orbital magnetic field.

Electron does not become coherent - two magnetic motions reversed.

Electron has two magnetic fields one inside other, orbital particle.

The Electron wraps one spin around the atom to become its orbit and proceeds to orbit in reverse of its particle spin.

Electron is opposing itself along both the tempic and magnetic fields.

Two magnetic fields in repulsion on one another.

Electrons field swells out because it is repelling itself.

Magnets are explained from this dual spin quality of Electrons.

They will only attract if both spins are aligned and two flows move through both the magnets.

Proton is not, it is in attraction with itself.

Its fields pull inwards in attraction and shrink.

Electrons - two magnetic fields – interaction.

One always connects with Proton in attraction.

One always repels from Proton.

Either state will transfer Torsion and is thus stable if there is some measure of polar alignment.

Proton nuclear flips become possible and are stable for this reason.

When the electron shell is in orbital attraction to Proton this is the low energy state.

Torsion moves to the whole shell as one torque on the orbital motion.

When in particle attraction this is the high energy state.

Torsion moves from Electron Particle to Protons field.

If we reach a quadrature alignment then torsion disconnects as Particle and orbital alignments are always in line for a particle and its orbit.

Neutron

Proton inside Electron. In the Neutron, Electron has lost its orbital spin, but still has its dual spin quality called half spin, one up, one down. It is at this point the particle becomes the diamagnetic field, with a Proton setting inside it, and a higher light speed constant.

Proton can bond inside Electron with alignment to either spin, but here they are more balanced so it sits at 90 degrees to become neutral. As its axis turns inside the Electron it will alter the dual spin of the Electron and becomes the Diamagnetic field, able to expand and contract at the greatest range its torsion which is normally balanced. The diamagnetic field will turn spin of the Neutron particles mass to 90 degrees of the magnetic field, disconnecting its torsion from the magnetic field and becoming almost Neutral util acted on by outside forces.

The presence of another Proton causes the strong force to pop out by influencing this balance.

Electron will bond to Neutron magnetically with almost no torsion forces present, as both have dual spin. Electrons will set right on the edge of coherence in opposition on one spin and attraction on the other. Proton will bond to Neutron with strong torsion forces and weak magnetic forces. Electron and Proton will interact magnetically and torsionally to create the sloshing stuff we call electronics, causing inductance.

Any imbalance in the magnetic fields between the Proton and the Electron shells will bring torsion out of the Neutron to counter the action causing the imbalance.

This creates reversed precession or drag in both outer shells and introduces the negative energy loop and the gravitational attractions. The Neutron acts to cancel all energy changes and pull them towards the Zero point. But this circular loop now creates a negative or reversed energy system inside the atom.

The torsion will flow to the Proton shell first because the Neutrons outer shell is an Electron. Dominant spin will always grab the Proton first and start to turn it towards the 90 degree spin of the Neutron. The torsion will now connect with one of the

Electrons torsional components and try to turn it as well creating a magnetic field to cancel what was induced and bring all back to Zero Point. It is the Electron on the outer shell of the Neutron that will be able to produce power full EM forces when the Neutron is acted on by either EM from the Electron shell, or Torsion from the Proton shell. One side of its dual spin will increase as the other decreases.

Over Unity:
[Breaking the atoms self regulating negative energy loop]

Scalar Coil

Using the atoms stabilizing control loop. Set an outside Electron magnetic field to counter or inhibit the Electrons torsion. Opposing magnetic fields.

Diamagnetic field at the Neutron will respond to strengthen it through the Protons torsion.

Proton will respond to meet the outside force in opposition.

If Proton succeeds it will cause the atom to receive torque or motion to its magnetic field at 90 degrees, so as to achieve balance.

To override this reverse energy balancing loop, we must set diamagnetic field against diamagnetic field such that the torsion connection is lowered or broken. Then we can tap one of the atomic spin forces directly to power our device. If the diamagnetic field expands to place torsion on the Proton far enough then any copper setting near it will receive this torsion and begin to respond, as long as the torsion is not balanced. Since the torsion is not directly coupled to the Electron shell but the Proton shell, the diamagnetic field will not effect it directly and should pass through. The wave produced by this will not have any electric or magnetic properties, and will not be visible on scopes or meters. It will be a torsion or tempic field, a one dimensional force field.

The diamagnetic field in its rest state is radiating a balanced Torsion force outwards at a higher light speed setting at the Zero point of our density. As it responds to balance the atom this is pushed off balance. If it can not succeed it will increase this force. The diamagnetic field will come all the way out of the atom and can be tapped directly using Copper atoms and the torsion field.

Two Atoms

If we are able to bring the diamagnetic field out of two copper atoms such that each is trying to move the electrons the opposite direction, they will meet and counter one another establishing a new balance between them. Neither atom will succeed because they are both opposing and the diamagnetic field will expand to a higher power level. Each atom now off balance in them selves, and yet the two forces countering equally between the atoms.

The diamagnetic field that we receive will now be expanded but balanced by two atoms, so at this point it is still not available to do any work.

To achieve a copper interaction we need to now drop one side from the expanded state, throwing the expanded field out of balance, and capture the other one as it accelerates back up to speed. This requires pulsing and coherent Proton torsion release. The two can be unbalanced in many ways, but pulsing will work best because it will act to lower the Proton to Electron torsional connection. We ideally want to pulse only one side while the other side is kept at the same balanced reference level.

We have many options for pulsing the fields in many devices but the principle is the same. Establish an enlarged diamagnetic field by setting the Electron forces into balanced opposition, then alter the balance between them with a pulsing mechanism. Capture the diamagnetic fields imbalanced state during the recovery time as a force that will move Protons into motional torsion at it strongest point by guiding this into Proton coherence.

The Balance Point

During the time that the Electron action is stuck or retarded, the Protons torsion will be highest and setting spinning in two opposite directions in each atom 90 degrees to the magnetic field. This will be effecting density. The torsion field has the greatest reach. These together are in alignment with the diamagnetic field which is setting them into motion. On the conscious side these tempic fields will always be coupled [density] but on the physical side they propagate through torsion between particles. So at the balance point we are pulling on the Aether directly. Tempic force vectors in opposition in the physical raise density in the Aether.

The Electric field will be lower as Electron motion is retarded.

Unbalancing the Flows

Pulsing cycle for diamagnetic field manipulation

Scalar coil pulsing technique -
Pulses must be syncronized to start together

As diamagnetic field apears at max - drop only one side of the coil
Torsion induced from colapse should be much higher

Roddins method was presented backwards from this?
Could be he is looking from the conscious side.

During an uneven release of the fields in balance, this torsion will shoot out at 90 degrees from the atoms magnetic field alignment, as well from its highest velocity side, at the same time it begins to add torsion to the Electron shell. There will be a surge of both the torsion field inwards slowing and the electric fields outwards accelerating, by leaving one electric field in place and dropping the other one, or even reversing it. Electric current will shoot up from stationary to a high current, and torsion will drop from its highest velocity to some lower balanced state. The energy is thus transferred between the electric and torsion fields in this manner.

Simple Apparatus:

Over Unity Scalar Coil Aparatus

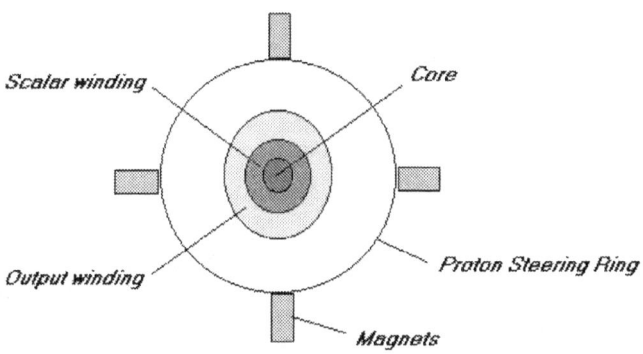

This action can be done with a bifillar coil with both halves of the coil fed separately. Wrapping the coil on bismuth or copper or iron will effect which field is more coupled to, on these releases. As the electric field surges with energy it will create a magnetic pulse in the iron from the bifillar coil, this effect will be weakest because of the distance reach of the magnetic field. As the torsion fields swing out of balance they will induce a field in Copper leading to electric flow if they are coherently steered. The electric field density will not be shooting out as far as the torsion fields, and the magnetic fields will have the shortest range.

135

The greatest field to capture is the torsion field and so another copper coil to capture this would provide the greatest energy transfer. This coil can then transfer the pulse into an iron core if desired and frequency is low enough. This pickup coil will receive a turning torsion field, as the Electron and Proton layers realign. It will turn from a 90 degree alignment back into a standard alignment with the bifillar coil. The strongest pickup would be expected to occur somewhere around 45 degrees to the magnetic field where it could capture the greatest swing of the strongest side of the field. A conical coil above or below the bifillar coils poles would probably be ideal to capture this pulse, although experiment is in order to determine the strongest alignments, also a 90 degree coil may work.

The weave wrapped coil will respond differently to this unbalanced pulse then a true bifillar coil if a center tap is placed on it so cannot be used for this.

Proton coherence

The scalar coil must be surrounded with a ring of magnets all pointing inwards, this will bring the non coherent Protons all into alignment so during the Torsion swing they will be coherent and all be present to induce current in an output winding setting around the scalar coil. This is not known for sure, but the magnetic fields will be cancelled at this point. The magnet polarity will determine which side of the coil to drop that will create the biggest pulse, because the magnets themselves will unbalance the torsion fields as with the RainMaker units.

If the magnets on the Proton steering ring are electro magnets, then it can be reversed on each cycle and output will become AC. Also pulsing must be reversed on the scalar coil, dropping the weakest side first. Steering ring can be reversed or even now pulsed to only be present near the peaks of torsion release and little energy may be necessary to accomplish this goal. The Protons must be aligned before the release of one side of the scalar coil, so this timing can be experimented with.

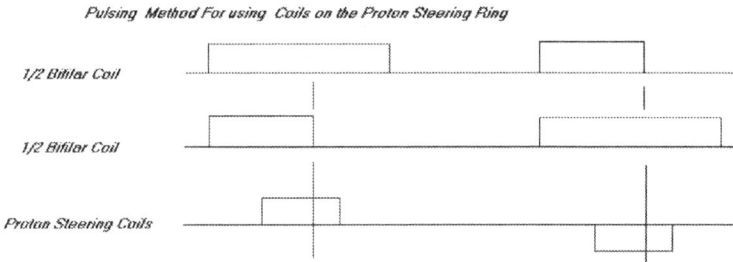

Pulsing Method For using Coils on the Proton Steering Ring

Steering Field must be present during Torsion release

For strongest outputs the coils can be wound together, 3 wires as one coil. Alternates would put the output coil inside on an iron core, since the torsion pulse will be coming from the scalar coil and not the iron there is no need to place it on the iron core unless a large magnetic pulse is desired as well. In this case the output coil can be shorted. Magnet strengths and ring size can be found experimentally for the weakest magnetic ring field to produce the largest output pulses.

Copper Thruster

Using this method on a copper core may lead to a torsion added to the core moving it through space in one direction with no opposing force.

A pulsed engine thruster. If torsion is raised both directions simultaneously, and then dropped on one side first from its highest velocity side, the torsion force will move through only one side of the coil and not both, thrusting away from one pole of the coil during the unbalanced time interval. The thrust may include a spin, so two set up together would balance to a directional thrust, powered by the atoms diamagnetic interaction.

Gravity Effects

Floyd Sweet noted gravity effects in his VTA units, and the reason may become apparent in this sort of device using scalar coils with an external magnetic steering field. As the Torsion is released and before it is fully engaged back to Electron regulation, the Precession drive at the Electron shell may be absent. If this can be regulated to last the maximum amount of the cycle then a proportional time will begin to appear releasing the gravity connections for an interval of the cycle. Although this will not lead to total weightlessness, it may produce an effect. If the torsion field were to completely collapse then this system would begin to fail, not being able to produce any output power as the torsion connection from Proton to Electron shell is totally severed. Driving the scalar coils to a higher level would be expected to produce higher gravity effects, as Proton tilt would be increased towards a quadrature alignment.

Dave L

The Field Forces

Tempic
Electric
Magnetic

If the field forces are studied in reverse then understanding may become evident.

Most get to the second and this is where our logic breaks down, and then the wild theories begin.

Wilbert Smith identified them for us, and stated all that is in the universe must comply.

Magnetic

This is the three dimensional force and corresponds to a volume in space. It is the shortest range force and field intensity drops off as a function of distance cubed. Much is already taught about magnetism, however I will now focus on what has not been taught at this point. The nature of field coherence.

Field coherence, is achieved at the Electron shell. When more then one Electron shell of a magnetic material comes into alignment at very close distance, then a single coherent field develops in the merging. The center or blotch wall of the field shifts. The four poles now merge into two poles, and the two atoms share the single magnetic field in a state of field coherence. This causes an expansion of the field much further outwards then normally present without field coherence.

Electron contains dual spin, the two spins repel one another, but in the orbital configuration they are not balanced, one is stretched around the atom and the other remains spinning oppositely. We end up with two magnetic fields overlapped and in opposition, one particle and one orbital, so the field balloons up and expands to the donut we are familiar with. One overpowers the other in this orbital mode, pushing one inside the other. This make magnets possible. The two opposing flows that are set up between magnets now interact diamagnetically to some degree because they are always in repulsion of one another. To establish field coherence they must merge to the point where both opposing flows are moving through both atoms.

A good demonstration is to take a long iron pipe and place a strong neo magnet inside one end. Take a compass down the pipe and watch the polarity of the field. The magnet will extend its own field to several inches and all the domains in the pipe will take on the same polar alignment over this distance becoming a coherent part of the one primary field, loosing their individual poles and blotch walls. Now at several inches down the pipe one will discover a reversal of the field and a new set of poles and blotch wall setting next to the other one flipped into repulsion. This point along the pipe where the poles flip is one reason the Electrons orbits are regulated by the complex diamagnetic field which has both attracting and repelling fields present at different distances. Where a small piece of metal is slowly moved away from a magnet, there is found this magical point of distance where its poles flip and it takes on a separate set of poles from the field it was inside and its own new blotch wall. The interaction causes the poles to flip into repulsion of the original field and interact with it. This is the true nature of Electrons setting in orbits of iron atoms.

When dealing with a coherent magnetic field, it can be treated as a single unit or a quantum field. As it breaks down it must be treated as individual interacting fields. Wilbert Smith stated, the point where both fields overlap over half their reality they become coherent, and can be treated as a single field.

At this point we realize that a great many explanation of resonance effects at the atomic level or the domain level can be dropped for an understand of coherent magnetic fields. What ever effects a part of the field effects the whole field for coherent Electron magnetic fields, which are the only ones emerging from the atom to any practical distance. The Protons magnetic fields are contained inside the atom and do not become coherent. The Protons coherence is with the Torsion fields.

Electric

This is the two dimensional force and corresponds to an area in space. Electric potential is stored in capacitors that are large flat plates very close to one another, with a dielectric insulator between them. Electric force field density drops off as a function of distance squared, so while it is a planar type force, its effective reach is further then magnetism. Electric force is not as flexible or cushioned as magnetism, and in order to establish a resonance a magnetic field is always necessary. Electrons must move between a capacitor and an inductor, you cannot resonate two pure capacitors or two voltages. It is an inflexible force.

Field coherence in the electric field is the Electron, however since it is in motion at slightly less then light speed there is no purely 2 dimensional force ever found standing alone we can call an electric field with no magnetic field present. Electrons can be treated as though they were static charge from the physical perception and in this view we create the electronics found today. Mentally picturing static charges and charge in motion as the magnetic field. The field of electronics is well established and needs no further discussion other then to point out that induction is not what we have perceived from the physical. Induction and the diamagnetic fields involve also Proton and Neutron force connections as well to operate, and the magnetic field is what operates the atom, together with the electric field in attraction.

The field of NMR is beginning to expose us to the Protons magnetic level of understanding.

Tempic

The physicist is supposed to explain the atomic models based on experimental observations, and yet none presently can explain why light is not bent by a magnetic field, as magnets fields are

141

bent. If Light were an EM wave having a magnetic component then it should be, however only strong gravity will bend light. In two opposing magnets we see the magnetic field is not only bent to the extreme but reaches out now further in opposition then without opposition, proving that magnetic fields do not cancel but expand along their tempic vector. Light is a tempic modulation, Gravity is an electro - tempic one.

The nature of the one dimensional force is not well comprehended today. It is related to motion and light speed.

If observed from the physical it is seen as Torsion, and the zero point of torsion is stillness or zero motion.

If viewed as though one were riding a beam of light, then the tempic field is ones true velocity of perception or C.

All atoms are light traveling in small circles, and this is the nature of the tempic field. It is found in light speed spin. Since light speed is not fixed, the tempic field determines, gravity, time flow rate, light speed velocity, and the forces that connect us all in the background Aether. The tempic field is normally a coherent field and connects us all together, forming the Aether. Altering the tempic field effects gravity, time flow rate, and light speed, it is the prime force. As a gradient it is sensed as a torsion field which is normally always coherent in our world but can be altered, and in many of our devices this can be directly sensed.

When one pulls on the background Aether to create a gravity field in space, it forms a well or sink in the Aether and alters the light speed constant, lowering it. The atomic motions of matter are seen to be very slightly slower then light speed of a photon and in the difference of velocity is found the nature of these connecting Aetheric forces.

The tempic force is a linear field density force. It drops off as a function of distance to the normal background level, and therefore has the greatest reach of any of the other forces, though is little perceived from the physical and most are unaware of its operation. Its Zero point lies in the Neutrons balance operating to pull all forces to the center of our density. The Zero point of our density is not "Zero" on the conscious side, and allows for density travel by shifting its Zero point.

142

Torsion is coupled through matter at the Proton shell and at this level it becomes a coherent field.

Torsion is coupled and transferred through a magnetic field. This can be seen from holding a magnet near a spinning copper cylinder, and then holding a nonmagnetic bolt near the same spinning cylinder. There is no other path present within the atom to transfer nuclear torsion but the Protons magnetic field. It is present in all matter and why each element has a unique NMR signature.

The Main Torsion center on the atom is the nucleus, and if there were no magnetic coupling between the nucleus of atoms then the torsion field would collapse and matter would not be coupled into the local time flow rate, it would disconnect from all outside references. This creates a different possible time sheer effect where a craft my disconnect from gravity and the local torsion field. This can be done by breaking the torsion link at any point, but inside the atoms torsion must disconnect between Proton and Electron shells. In a diamagnetic liquid it may happen between molecules. Or it may be done psychically as with party levitation at the conscious level of the diamagnetic field.

The Mirror Symmetry of Light

Forces viewed from the physical reference appear to interact opposite as they do if viewed from the perception that we are the light or riding the light beam of the field force. This is from the Spiritual perception.

An example is two magnets placed into opposition. Where we would expect opposing forces to cancel, in the physical we see two magnets move together and stop one another when the equal force of repulsion and compression is achieved on the physical perception side. However the actual interaction creates an expansion of the field on the invisible side, setting between them along each ones tempic spin motions. The tempic vectors representing only two of their volume expand, rather then canceling at the level of their field force. The invisible magnetic forces do not cancel but increase in volume of two dimensions while compressing the third. The three dimensional force is not lost only shifted into another tempic vector component of its field in space. The field balloons outwards increasing its physical reach in space. Either space is now altered or the velocity of the magnetic field. If we reduce this to tempic vector interactions we discover that in the magnetic field, opposing spin vectors cause an expansion rather then a reduction of their physical dimensions. Since the same energy is now reaching further, one must become aware of the possibility that the fields velocity may now be higher.

When observing the tempic field, the same is true, only the elastic shifting between dimensions is lost. If we place two oppositely spinning gyros near enough to exchange torsion, the field interaction between them increases the light speed constant of the tempic field expanding time itself. Time flow rate is altered in the upwards direction. The nature of the one dimensional force as perceived from the light perception side. If we now add a 90 degree torque to the axle of the oppositely spinning gyros they begin to exchange Tempic fields at a greater field intensity and the interaction becomes stronger. The Torsion fields expands the light speed constant operating between them. This force spreads out as a function of linear distance to form a well in Space against the background field of 3rd density. This is the model of forced precession and can be applied either at the device level or at the atomic level in the atoms of the AG materials.

The Model of Light

Light or a photon was intuitively discovered in the Creation document, however now it becomes more obvious. Light is not Electro Magnetic as we are familiar with it. It is a Tempic vibration or loop which has no electric or magnetic field until it hits an atom able to absorb it into its electron shell adding back the EM components of its wave. Light is not bent by a magnetic field or an electric field because it does not have one in itself to be interactive with, it is only bent in a gravity field where we see stars bending the light from objects passing behind them. It is the smallest pure tempic field vibration we normally encounter. Much like the torsion waves we see Protons emitting in the scalar coils, they can not be measured or detected using EM sensitive equipment, until they hit a system designed to move the energy back into the normal EM fields. Light does not loose energy as it propagates, the value of Planks constant shows us that it is a quantum energy packet that does not experience any loss as it travels. Torsion waves also exhibit this quality.

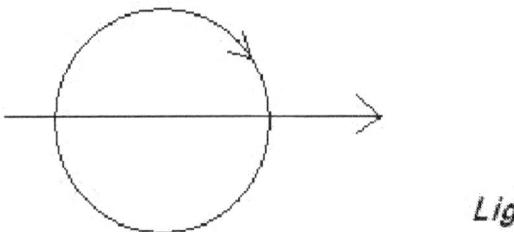

Light

Light can be polarized, it only fits one direction through a narrow slit smaller then its wavelength. It is a flat loop traveling along its edge. This is the same model for tempic waves or torsion fields. The longitudinal wave is moving through the tempic field not the EM fields, an important distinction.

Light becomes an interference pattern only after it interacts, study the two slit experiment to see how this operates. Light is operating inside the one dimensional force at its most basic level of primary spin. It is an alteration of its velocity back and forth as it moves through space. Wilbert Smith got the answer to, what is light? "It simply is". It is the first possible vibration operating on the prime force of nature the tempic field. Thus the torsion field is also this type of manifestation, completely void of EM. This is why scalar waves may pass right through solid objects such a Faraday cages, if these waves are not responsive to the wavelengths of the matter.

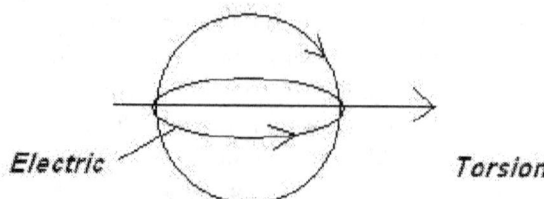

The tempic field receiver. Two coils placed in quadrature will rebuild the electric component of a torsion wave. Only after the Electric field is reestablished will a magnetic field form, from its movement, and we discover all sit at 90 degrees to one another.

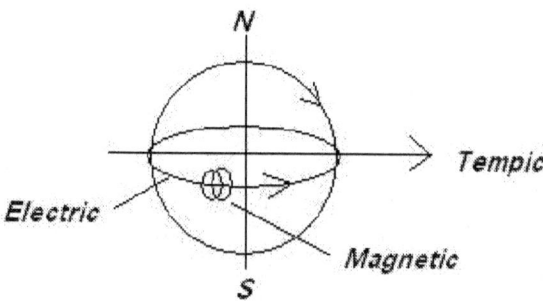

The donut now forms around the electric field in motion. Here we end up with the familiar Electric Magnetic and Motion vectors appearing in Electronics, only we have created them from the prime force outwards and included their correct spin angles to explain light polarization and torsion fields. This model shows us what we have come to call Tempic field, Torsion field, Motion in electric motors, and Light photons are all manifestations of the same basic force, the prime field of nature. Although they all may interact with EM under correct conditions they are not of themselves EM.

It should become apparent due to the nature of matter, that if we could succeed at pushing matter up to its local light speed, it would now be moving along its torsion vector equally as fast as it is moving in its circle pattern, and the circles would never complete as spin. It would become spread out into a stream of energy with nothing that is recognizable to anyone observing it. The solution is to increase its' light speed constant so it can now make more rotations, moving it through density, this requires a tempic field manipulation.

Gravity

Gravity was explained by Wilbert Smith as well, but not comprehensible until one grasps the tempic field. Gravity and the Electric field are composed of the same two vector forces, tempic and electric fields setting at 90 degrees to one another, and thus both offer a distance squared force. Electric force projects along its electric vector which has a gradient or a modulation and carries a constant tempic field along with it that is relatively unchanging at 90 degrees. Gravity has a relatively constant electric field but its tempic field contains a gradient or modulation, and thus it is able to effect light which travels the tempic field. The Electric field gradient is not known to alter the path of light and therein lies the difference.

Collapsing the Torsion Field

When we observe the work of John Hutchison, levitating a cannon ball using 5 EM instruments, we realize that the forces of the tempic field can be altered. What is happening may not be clear to most and seems like magic to be feared, but the explanation is really very straight forwards from the field forces models already commonly in place if one can make the leap to the one dimensional field force, or Torsion to realize its true nature.

Study of the atomic system reveals the necessary facts:

The major weight and mass of an atom lies at the nucleus and is moving at near light speeds in a state of spin. The rest state of this spin is C or light speed, the zero point. This can be viewed as a torsion force from the physical side, and torsion can be used

147

to make electric motors spin up. However this involves a 90 degree shift of the motion of Electrons into the motion of Copper atoms. The effect is based on spin and originates inside the copper atoms interaction of two spin forces operating at 90 degrees. This is the natural positioning of the magnetic and diamagnetic fields as they are repelled from one another.

In Copper we see it is magnetic at the Proton layer and not at the Electron layer. Placed in a moving magnetic field induction causes the diamagnetic field to expand outwards becoming a strongly repulsive force. It is the diamagnetic field that is spinning at 90 degrees and why the force is transferred into another motional vector. The diamagnetic force is always present to offer a stronger force always pushing back but not in the same direction of spin.

Torsion is transmitted or coupled through a magnetic field, remove the field and the torsion stops being transferred. This is the nature of field coherence for the torsion field and how it can be collapsed. Torsion moves between atoms through the magnetic field of the Proton, atom to atom, and over distance becomes the field with the greatest reach outside of matter. It hops from Proton to Electron to Proton.

To collapse the torsion field in matter from the physical side, requires manipulating the Protons such that they can no longer form into their natural isotope chains of alignment and exchange Torsion over time through the Electron shell. Field coherence will drop away and the torsion field collapse will disconnect matter from all tempic connections in the outside frames of reference. This includes the gravity, and time constants which operate through this link.

In party levitation we see 4 people using the diamagnetic field to consciously collapse the Torsion field coherence in the center person. The person may experience conscious time disorientation and weightlessness, and is easily lifted with only a couple fingers from each person. Due to the conscious nature of the diamagnetic field, it can effect torsion from the light speed perception where consciousness is truly located, however on the physical perception we must resort to alignment of the Protons magnetic fields.

Two Methods Previously Identified

Hutchison

Hit the Protons with so many EM pulses in a noise or at random rates so as to keep the Protons in a state of Chaos, with few magnetic fields crossing long enough to exchange their torsion. The torsion field looses its coherence and crashes inwards like a magnetic coil being de energized. The EM forces of each atom continue to hold matter together. Although with this system we see that often even steel may loose molecular cohesion and begin to fall apart. This means the torsion became non coherent inside the atoms as well rather then merely shrunken to the atoms size. The unpredictable nature of this method, although invaluable in proving to the world it is possible, is not suitable for common use, nor is it safe to be around.

An offshoot of this would be to solve the problem of creating an interference pattern to dislodge the magnetic alignment of Protons into a randomness without effecting the Electron shell in the process which operates in the microwave band. The correct pattern in the 1 to 20MHz regions, or at least well below 1Ghz. This is a problem for the mathematicians and physicists to solve but would lead to a device that could be placed against an object and cause it to become weightless just as with party levitation.

Otis Carr

Coaxing the Protons into an ordered phase alignment where synchronization of atomic precession fall into 180 phase shift to their normal state.

This method will break the torsion link by achieving a new synchronization of all the Protons to precess together rather then forming the normal 45 degrees alignment reversals. It involves also expanding the diamagnetic field to lower the magnetic attraction.

149

Combined with Tempic field Torsion interaction this can be used to effect the light speed constant. The first step is collapsing the Torsion field in matter and disconnecting from local time and gravity.

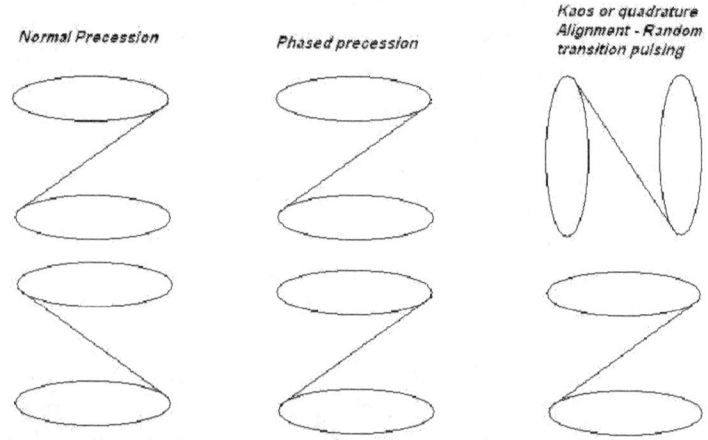

Coherent Torsion Release by phased precession

If the poles of the Protons fields can never come into alignment then the torsion field can never be transfered between atoms in matter and the Torsion field will colapse.

Two magnetic fields, one steady and one pulsed setting at 45 degrees to one another, should be able to slowly synchronize the Proton motions of a Copper mass into a reverse phase precession at the isotope line, if there is a strong diamagnetic field present.

Other methods have also already been identified to accomplish this and with a firm grasp of why we are using them, the fear factor should be removed, and progress accelerated in this direction. The key to the Anti Gravity devices lies in collapsing the coherence of the primary field force. The tempic field, as it manifests as Torsion and is transferred at the Proton layer.

With a contracted Torsion field in place both momentum, or resistance to motion, and motion through the background field of space is lowered. Thus the high velocities we see appearing in the Searl disc that should by rights shoot the rollers off the machine, where the magnetic field holding them meets the centrifugal force pushing them outwards. Physical laws seemingly being broken, yet with a collapsed torsion field this becomes possible.

Otis Carr's spinning cones also reaching incredible velocities that no flywheel could possibly reach if the torsion forces were at full strength. To reach an equivalent RPM of the earths rotational energy, is much faster then any gyro can spin and remain together if centrifugal force is normally present in the gyro.

The solution does not lie in finding new super alloys to stand these higher forces, and in stronger motors to spin them, but in reducing the forces themselves.

We have a very good aptitude for the torsion forces from design of motors and large flywheels, we have absolutely no concept for operating in a mode where torsion is reduced or eliminated. This will be coming! Some of the side effects for one caught in such a field, confusion, time disorientation, weightlessness.

Diamagnetic Field

The diamagnetic field is a complex force. It is made from a very unique combination of the above root forces and labeled a force, however it is not a single natural force. It is reactive. The Neutron is a complex structure consisting of one Proton setting inside one Electron. This means it reacts differently to Protons then it does to Electrons. Torsion may move directly outwards to a Protons magnetic field, but it will not move directly to an Electron magnetic field, it must first move through the Protons to get to the Electron shell. If we hit a diamagnetic field with an Electron magnetic field, what we get back is a reversed magnetic field and a torsion less connection. If we hit the diamagnetic field with a Proton magnetic field we get back torsion directly. In this sense calling the diamagnetic field a force is not totally accurate however it does convey the meaning of a composite field that interacts uniquely. Electrons dual spin is in balance at the Neutron.

151

Motional Considerations

The diamagnetic field is in motion. Its torsional poles are spinning along the equator of the magnetic field in small light speed circles as they turn.

We can think of the fields inside atoms as shells or spheres that turn inside one another and interact. The magnetic fields poles are normally setting pretty stationary with respect to light speed. It precesses its poles in small circles top and bottom because of the atoms natural negative energy drag. In the Electron shell at a microwave frequency. In the Proton shell at an RF frequency, but both are in alignment and both are precessing.

The diamagnetic field sets inside these and spins at 90 degrees, very fast. It also rotates its poles along the equator of the magnetic field. As we begin to interact with the atom to cause an expansion of the diamagnetic field, it interacts with electrons to propel them at 90 degrees to the magnetic field using opposing EM. More importantly is slows down its spin and this spin moves into the magnetic field as motion. As we increase the spin of the magnetic field at low speeds in conductors it will hurl Electrons. As this continues, the magnetic field becomes the one in motion and the diamagnetic field becomes the one that becomes stationary. Neither can be stationary at the same time with respect to light speed velocities.

If this process continues we end up with an atom spinning around its diamagnetic poles which become stationary and coherent.

The magnetic field is now the one in motion moving towards light speeds, and its poles nearly disappear into a blur.

This describes a world where all magnetic fields disappear into non coherence, and the diamagnetic field becomes the dominant force coherently moving through matter.

Torsion radiating and coupling through the magnetic fields of the atom is disconnected, and the torsion coherence collapses.

The torsion radiating from the diamagnetic field is an unknown and we do not know how it will interact with other atoms in the same state. But due to the nature of the diamagnetic field to cancel most all its magnetic field, it is not likely that normal matter will be able to sustain a coherent Torsion field at any great distance.

The answer to control of the torsion field and the tempic field would lie in the amount we shift the fields into the diamagnetic state and away from the magnetic state. The conscious interaction would also increase.

However as we see in the water, it is only necessary to begin to block the magnetic field link between atoms to disconnect the torsional link. As the diamagnetic field becomes present at a strong enough level in the water, setting the molecules into motion moves its field all directions over time and begins to block the normal magnetic field. As water is slowly squeezed out of warm air hitting cold air, the first molecules to appear do not fall to the ground. They have not coupled to enough other molecules to form a coherent torsion field. It is not until the waters natural charge collects enough atoms to form a drop that gravity begins to couple to them as one coherent torsion field interacting with it.

Torsion Bonding

This may be a good place to introduce a new term to science, based on my experiment 1 and Kosols water experiment. Torsion Bonding between atoms.

As the coherent Torsion field looses its' torsion bonding between atoms it retracts within itself and becomes an independent field to itself, much like one domain of an iron core does when the magnetic field is shut off. We have not destroyed the source of the field or damaged the atoms in any way. We have simply manipulated their natural fields to produce one state of matter, with either no or very little torsion bonding between atoms. The atoms torsion is still setting there inside it at the nucleus. The field reach of the non coherent state is simply far smaller.

Torsion versus Light speed spin [Aether] [Merging the Spiritual and Physical sciences]

It takes both a physical science comprehension and a Spiritual comprehension to grasp the connection between Torsion and Light Speed. Torsion, grounded in the physical plane, and Tempic field, grounded in The Light. The scientists are almost ready to accept the Aether, grasping for explanations at this point, because it is necessary to proceed, but they do not realize this will move them into the conscious planes where "We are the Light" and things operate differently, but it is the true connection between all things and the first words out of the mouths of the spiritual adepts.

On the tempic field, to understand you need to develop a concept for time flow rate and where it originates. Einstein shows us the gravity effects are connected directly to light speed. Einstein however claimed that linear space is warped by gravity. Wilbert Smith shows us that light speed itself is warped instead and a lower velocity creates gravity. Indeed this is obvious when one considers the velocity of the atomic particles are some .9999999....5 of light speed. They are slightly lower.

A gravity well is created from slowing light speed on a section of the conscious planes or the Aether, the Light we are all connected within. However it propagates as the precession motion of atomic particles moving through the Torsion field on the physical plane. While we can alter the coupling in the torsion connection we cannot ever disconnect from the Conscious side where we are all connected in the Aether, the Light realms. Even a device disconnected on the physical plane side, is still coupled to the Mental and Astral plane, and this is why it can be controlled through conscious means. These planes of Light are the Aether that connects us all. They are the worlds of Light experienced by the mediators of the past and present and are very real.

Light speed and gravity, two field forces, do not originate on the physical plane, as do electric and magnetic, they are coupled to the Light where our conscious side lives. Party levitation shows us this is true without a doubt by removing gravity from that side.

Gravity is a 2 D force created from the precession motions of the atomic particles modulating the tempic field. Classic Wilbert Smith - [Gravity is the precession motion of atomic particles each setting in different light speed references connecting through the tempic field]. however - in my revelations I discovered that gravity is the push of God force outwards into the lower planes, and this force crosses over from the worlds of light into the physical to become a pull towards matter. This force is recognized to accomplish this goal in the physical by making it very hard to leave the earth keeping various cultures separated and spread out in the universe. Shifting density, will effect gravities pull, but dropping the torsion link will disconnect it only on the physical side.

The gravity field flows through the tempic field [it is coupled to light speed and originates there]. And the tempic field is connected between all physical bodies by the coherent torsion fields linked through matter as Torsion between the nucleus of atoms.

The coherent field reaches out a great distance, but on the conscious side the tempic field is coherent everywhere and cannot be disconnected. It can only be "pulled on" creating invisible gravity wells, or the opposite.

As we disconnect the torsion fields in the physical plane we are still connected to the tempic field on the conscious planes, because we are all connected in the light. This is what I would call Light comprehension. It is the merging of science and spirituality. We disconnect the physical side but mind can still function to move us through the conscious side because it is more real then the physical side. Without this connection we would be lost to the universe.

This was shown to me in my first scalar crystal experiment, where a conscious link of 5 people on the mental plane allowed me to re orientate myself and get home.

Switching between the world of the spiritual and the physical offer many paradoxes because as we move back into our light bodies our perception becomes the light and no longer the physical. Things can appear reversed from this perception and why it is so hard to channel actual devices.

Life - Death becomes Separateness - Awakening, Torsion becomes light speed in the Aether - Our soul moves from the flesh to the Light - etc.

While in both we can become aware of both and operate in both. To travel in a ship disconnected from the coherent Torsion field requires Spiritual abilities be developed. However as the Yogis have stated, to change our level of vibration we need to be in the physical. This gives us access to the control vectors of the control fabric where we can alter density.

If there are beings out there traveling this way I would expect they are all enlightened beings.

On Time

Can there be time flowing backwards, or conjugate waves moving opposite directions through time and still passing through the present moment, to produce physical forces in the now?

There are many theories offering reverse time flow and using math to represent this. The math may make sense but the basic perception is not comprehensible when applied to the physical universe, or the conscious one.

The Yogi Guru tells us there is only the "now." All else is illusion in our minds. Wilbert Smiths, "boys up stairs" tell us "the Light simply is." However, Light speed is not fixed.

The "cold energy" is a tempic field gradient, a torsion force, or alteration of the time flow rate, not a reversal of its velocity to less then "nothing at all". It is beyond the yin and yang and operates above this with no opposite. It is the third element, or the "One." The **Yin** the **Yang** and the **One**, are the elements of the TAO. Only two may be reversing, the third one "**simply is**".

Time [conscious] or tempic field [light speed motion] is the correct dimension we need to add to the equations, but we need to discover its true nature outside the yin and yang, for our physical science perception. It is the primary force and does not cancel as does the magnetic and electric fields. As we cancel the EM fields the tempic field expands, it does not reduce to Zero and then reverse to a negative value, but does the opposite. It increases its velocity. Thus as the light moves out into the universe and becomes matter, the time flow rate lowers and time can be fully experienced.

It is good to recognize that time is involved in the modulation of the tempic field as altered light velocities, but if the tempic field stops or reverses then so does light speed, because this is the zero point of motion. These are torsion waves, and photons, small variances in the tempic field flow rate within the Aether of our density. How does light move slower then "stopped" or in a negative velocity? This is a conscious paradox, and must be recognized as a confusing of Spiritual and physical "perceptions." We are the light, if it stops then we stop with it, total spiritual destruction of all our vibrations within its primary field. There is only the present moment in motion, and the Light simply is. This is our limitation as light beings, and we do not exist outside the Light or consciously outside time.

As the physical world is rooted in motion the conscious realm is rooted in time, and storing memory in sequences. Reversing time on the conscious planes removes collective memory and lowers comprehension. Reversing light speed to a negative value reverses consciousness with it. If Source chooses to do this and run the universe in reverse then time flows backwards and all is forgotten. Conscious beings thus live their existence in reverse. This is not the observed nature of the universe.

Much confusion is present if one tries to create a model for something that is no longer present. A scalar wave is not a wave in the EM field but in the tempic field after the Electron layer has lost its EM function by canceling it. No model is necessary to represent it as a canceled EM wave moving backwards in time.It moves into a different medium, the Protons layer, and operates differently while there as a torsion force which can be felt and experienced in a different way.

The confusion is resolved in an understanding of the tempic field, the Light, and the nature of consciousness within the primary force.

157

Dave L
1 / 24 / 2007
David Lowrance
Kosol Ouch
c_s_s_p group
Kosol_Core_Tech

The Crystal Laser Coil
Invented: 04.04.07

Why not using light for powering a crystal? It will emit a gentle field without any EMF effects.

I was really astonished about the effect of this kind of coil on the crystal. It can easily be felt with palm over the top of the crystal.

Using a 05,mW laser-diode with controller and an optical fibreglass wire from advanced digital audio equipment, we will get a possibility to power up any quartz crystal by winding a fibreglass coil around the crystal.

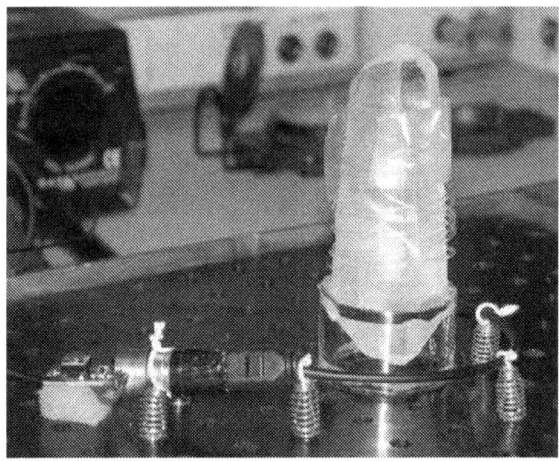

Below, the laser diode with the controller unit, connected to the fibreglass strand.

159

If some bismuth crystals are added around the coiled crystal, the field will get stronger above the setup.

The Laser – Sphere

Photonic crystal programming technology could be a more advanced way in future than using coils and electric frequencies. Using multicolor diodes (as I tested too), fibreglass coils and connection to vibrational datas offers a new spectrum in healing and braintuning - a next step in border science.

The night shot without flashlight shows what's really going on. The energy can even be felt on the picture.

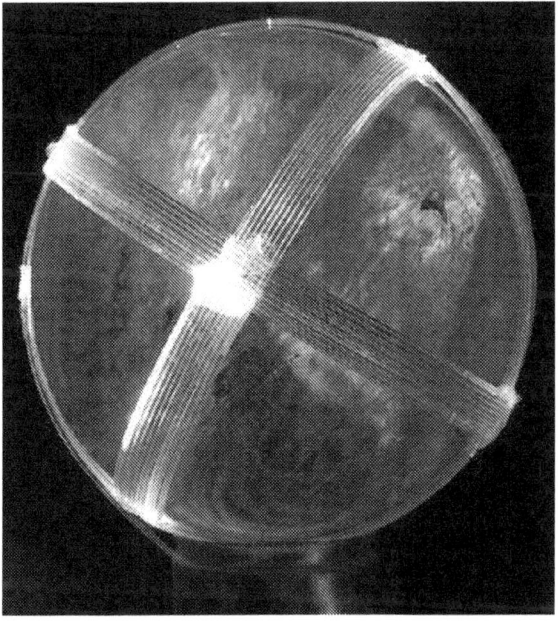

The Scalar Space-Time Chi Converter

The integration of my new wooden modulator unit into a totally new setup of radionic machine was the consequence of past experiences with moebius coils covering transmitter crystals, driven by frequency generators.

As I found they did work, but I always had some nasty side-effects detected with this coils and the generators on the operator himself.

As some other experimenters found ways to use new arrangements of coils to tap into existing tempic fields, new power sources could be detected to use as carrier for the vibrational data to be sent to a specified receiving person or even a vortex, created by diamagnetic devices as f.e. the RainMaker unit.

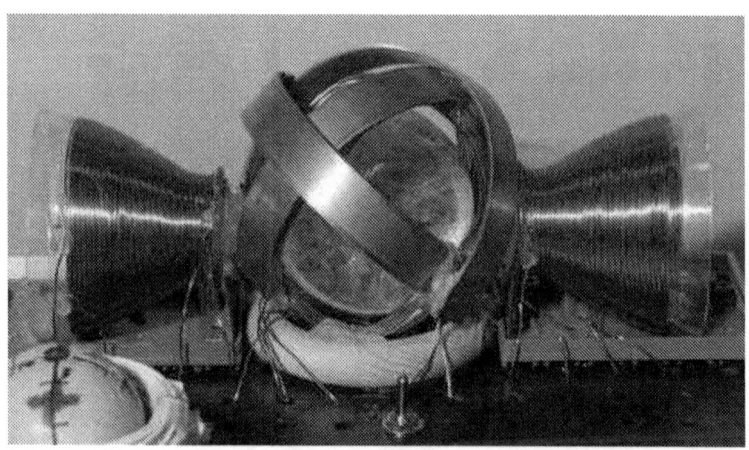

For some time I now use some special rings with their "Sacred Cubit" length of 20.6 inches.

The Sacred Cubit is based on measurements from the King's Chamber in the Great Pyramid of Gizeh. It has a natural harmonic constant in both length and natural resonant frequency that relates to the speed of light in free space, or 144,000 geodetic miles-per-second.

For the coils in this new device I used in all wirings the Cubit or the 4th overoctave of it. Therefore the wire length is 27.46 ft.

I found a gentle field emanating from the coils made in this way. Using a compression sphere with three of this coils aligned in right angle each other around a crystal orb, I could get wonderful relaxing fields emanating from my connected RainMaker device without using any electronic generators as usual.

So, the way led to the compression sphere as the reactor unit inside the new Chi-Converter setup. Left and right to this sphere I aligned a conical bell to get closer to the 45deg. angle for in/out streaming of the scalar fields. As transmitters I use double-ended quartz crystals with the "+" pole pointing direction outside of the whole arrangement. The crystal orb inside the compression sphere can be rotated for adaption to various setups of aligning the crystalline grid inside it to the emanating fields.

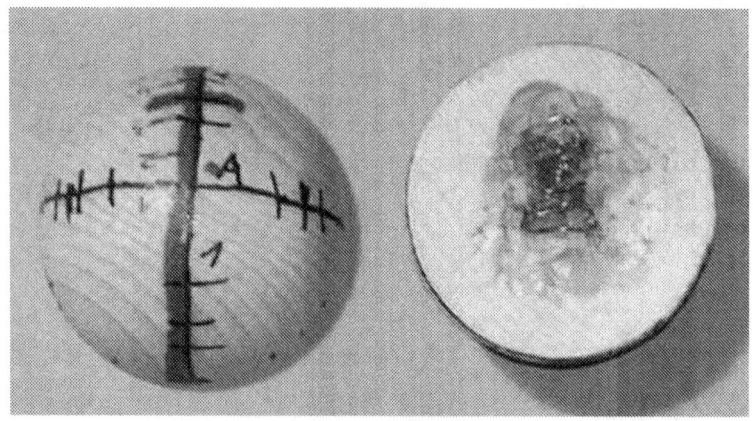

A view inside the modulator orb shows the double ended quartz crystal with caduceus coil. If this orb is placed inside another EM cancelling coil, there are in each angle different interactions to the overall scalar fields of the circuit. More details are shown in the "Chi Modulator" chapter.

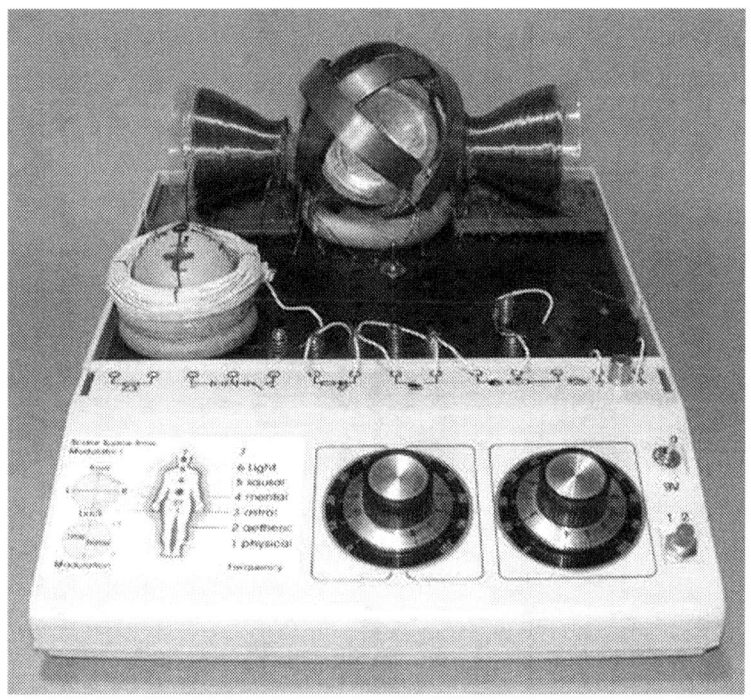

The whole device together with the modulator unit for giving interaction to the Chi fields. The modulator was the last discovery, which closed the connection to a universal vibrational machine for input and output, mostly using now scalar fields directly for tuning.

As I found, the crystal grid is important in orientation inside scalar fields, I got another tuning possibility - just rotating the crystal orb for specific output changes.

The other is the **form-ray-matrix** between congruent shapes - here we have too a very close interaction inside the device between the two orbs - as if they have direct contact without wires connected to them.

By programming the crystal inside the modulator, some various ways to use the device can be managed.

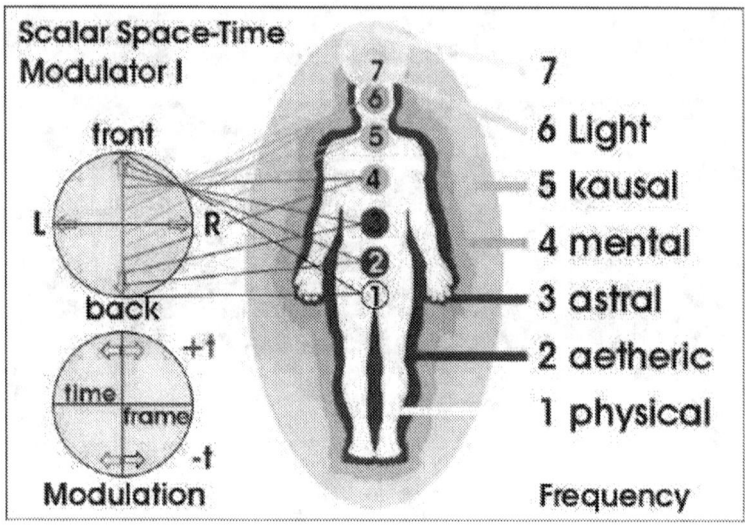

Above for example a chart as a way to interact to the human subtle chakras and by the capacitor to the subtle bodies simultaneously.

More and more I realized while running the device, that it has functionality of a portal to various dimensional levels...

I must point out, that in this state of development, the device is only for experimental purposes and use.

I do not claim any healing facilities for treatments.

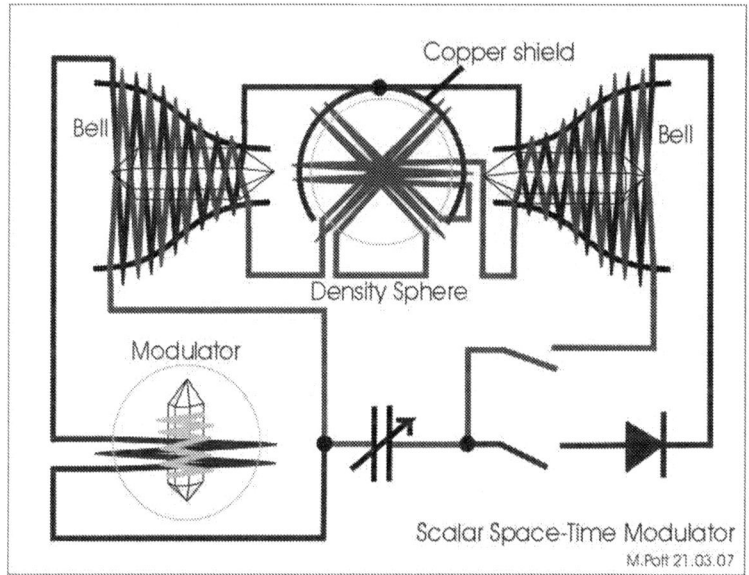

Above: the way of connecting the parts of the device.

The bell coils can be wired with one or two layers, each with separate leads and one in clockwise (red), the other in counterclockwise (blue) winding patterns. The density sphere has three coils around, all with 90 deg. angle between them. Some switches are necessary to shut down the device. I used 0,6mm magnet wire to wind the coils. If using stronger wire, the power will raise and it will need to separate all coil connections for shutdown!

The capacitor regulates some resonance points, that can be used for subtle energy research.

The copper shields are connected to the blue circuit. They consists of 1mm copper, covering each coil on the density sphere separately but with common wire to the blue circuit.

One Potentiometer (about 47 KOhm) can be inserted between the diode and the blue switch for controlling the general output.

All wirings are (20,6" x 16) /12 = 27,46 ft for the coils, except the small coil inside the modulator with 20,6" length

The 3D-Scalar Modulator Unit

This totally new setup I invented as a vibrational scalar tuning unit, using only crystal and coil interaction with a wooden sphere to control incident angle for tempic field coupling to the aequatorial cancelling coil around the sphere. No wires are needed between the sphere and the outer coil.

By this 3-axis movement possibility many vectors can be used and set on one control unit. Rolling the sphere in all direction, it always makes different interaction to the surrounding cancelling coil.

I see this as a possibility to replace all the lots of single-dimensional potis/switches used in the conventional tuning boxes. Only two of this new tuners are needed - one for the trend, one for the goal.

The wooden sphere was built by gluing together two half spheres, before gluing together hollowed for inserting a caduceus coiled double-ended quartz crystal.

As such devices are mind-interactive, we can use several programs for tuning. Without any mind programming, I found that the rotation gives some feedback inside the subtle bodies of a person. Rotating the sphere to front of my viewpoint, I got interaction with body-front side chakra positions, according to the rotation angle. Reverse rotation towards myself showed interaction to the back position of the chakras. Inclination to left or right with back/forward movement together could make resonance in the meridian system left/right side of the body. If turning the sphere up to 90 deg. right or left as a new start point for tuning can alter f.e. the timeframe of the output. This means you can send you some scalar hotspots into the future to arrive to the goal.

The wire length for the caduceus coil and the aequatorial coil should be in a resonant octave coupling, for example 1:16.

In my setup I used the Sacred Cubit length of 20,6" for the inner coil and 27,46ft. for the outer coil.

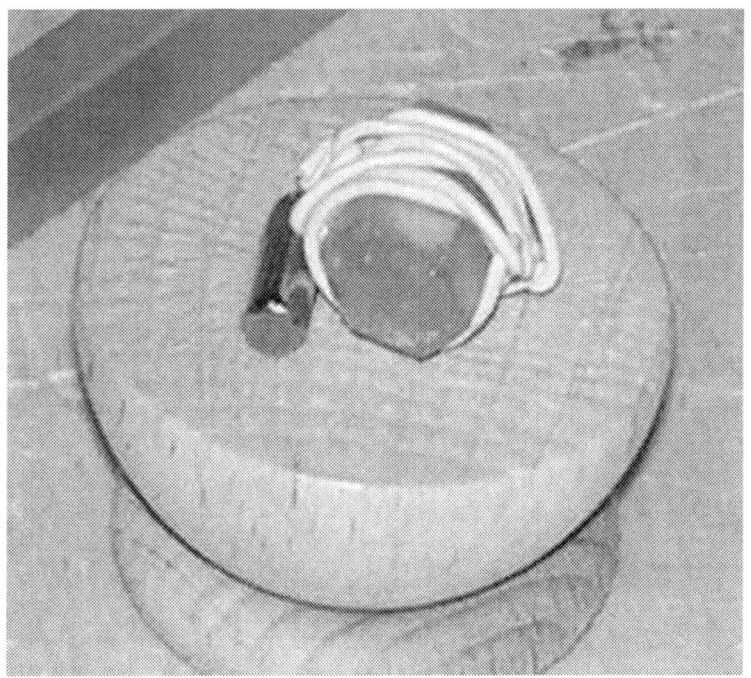

A half wooden sphere with a coiled crystal. This crystal is placed inside the hollowed wooden sphere - using better coil as shown and no magnet. The half spheres are glued together, wearing the crystal inside. The "+" tip of the crystal is the upmost point of the sphere in neutral position.

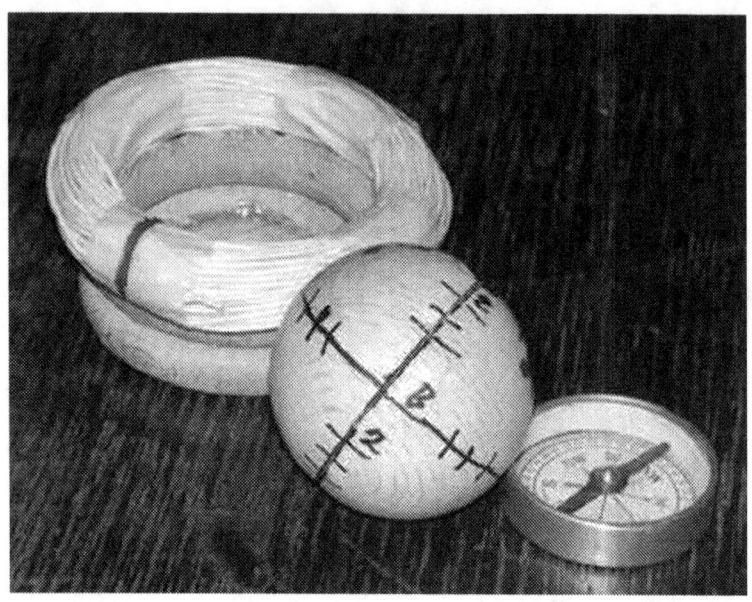

The finished sphere with measuring lines on it. Behind it the aequatorial coil - glued on wooden rings. Best work I found while the operator is facing to north.

The Tuning Unit in Use

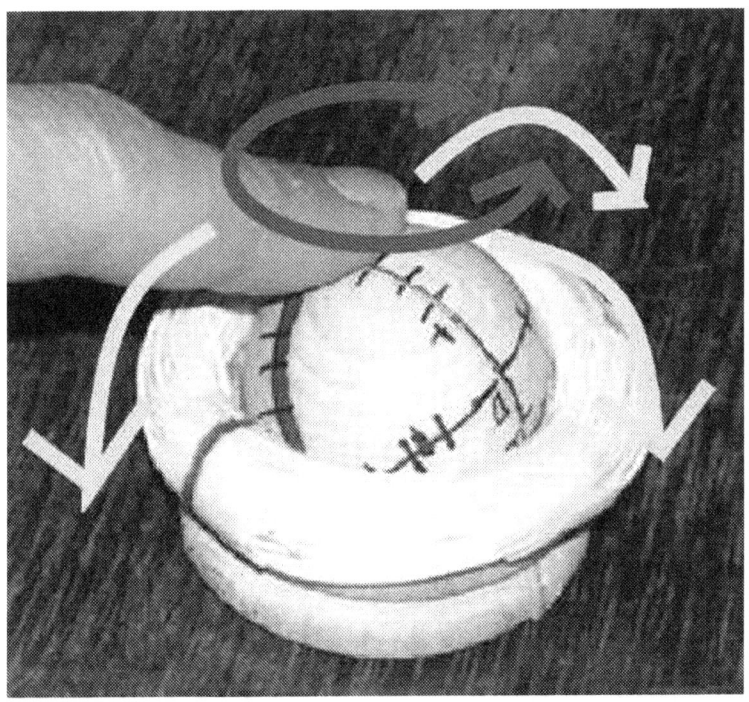

Arrows show the tuning front, back, left, right from the starting point with "+" tip of the crystal pointing upwards.

Arrow shows the possibility to rotate the orb into another angle while the crystal tip is still pointing upwards. This lets open another programming frame.

This modulator can be inserted in existing vibrational tuning units by the outer coils connections serial to the stick pad f.e. or near the output coils of a device.

The Full Spectrum Water - Broadcasting with the Rainmaker

An important task for those lucky ones having a RainMaker for their own could be, to help the nature and all beings in their surrounding area with the super quality of full spectrum water, broadcasted by the RainMaker device.

171

Now I made it very easy to proceed in this way to raise water quality all around us by the use of informational input of this holy waters into the vortex of the RainMaker.

I've tested this method for two months now, and can say it works.

You only need some additional parts and the water itself, an empty film box, a wooden plate, some wire for the coils and a test tube.

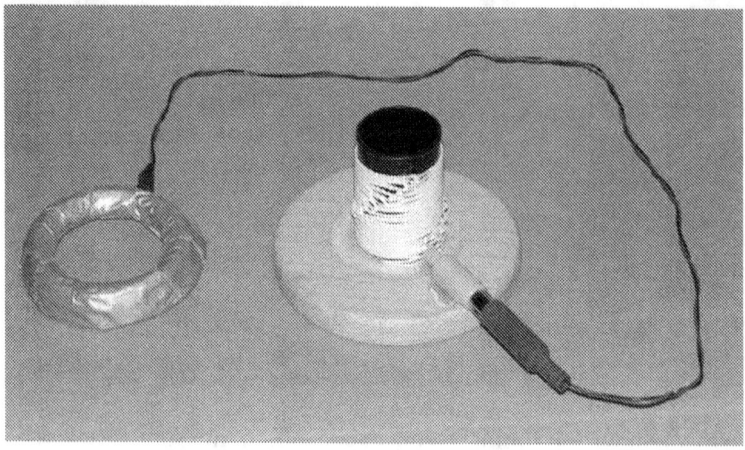

Just wind a caduceus coil around the film box and glue it onto the wooden plate. If you want, make the nodes on the wire build a right turning spiral. By this "hotspots" are running all around the box.

Then make a second coil in the same way as the existing ferrite coil of the RainMaker, This new coil will then sit on top of the ferrite yoke but below the crystal orb. Connect both together and you can start. The whole new setup will look like this:

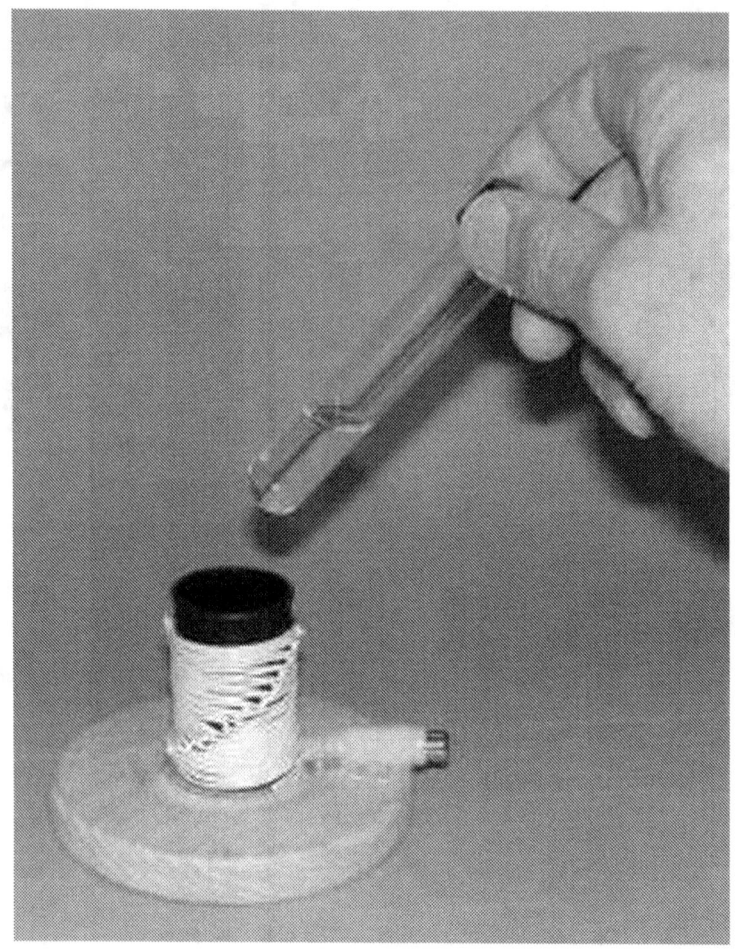

For the water you need one or more test tubes, best with dense closure.

Only some drops from the stock bottles of the waters are needed to inform a normal water quickly.

As some persons now tested the effect while inserting the test tube into the box, all recognized immediately the vibrational change of the RainMakers vortex. This means, that all water molecules inside our bodies react directly to the vibrational information of this water. The water molecules begin to arrange themselves in a new atomic way and this will create super comfort.

The waters of this special founts have been tested with spectro analysers and the result was, that each color was represented inside. It was a proof of the legend, that this waters can heal.

The effect to the environment is the next benefit of this new setup: the vortex of the RainMaker does inform all the water molecules inside his range - means clouds, rivers, plants etc.

Then the rain will inform in a chain of reaction all other water molecules on its way on earth.

We will get full spectrum rain and waters all around.

To get this water, you should check your local sellers in your country. Often it can be purchased in esoteric related stores or by ordering via internet.

There are too some books available about the spectrum tests and their results.

The Japanese researcher Dr. Masaru Emoto made tests with the waters too. He let grew snow crystals from them, which show with their beauty the quality of the waters.

Following a list of the fount waters, that are in the set I bought in Germany:

Fatima, Lourdes, Montichiari, Santa Maria alla Fontana, Medjugorje, San Damiano and Efeso

Technical Overview RainMaker 1 Device

Rainmaker 1 - Observed Process of Operation - Physical Plane Perception

For those who are not familiar with the basics of magnetism as it operates in the materials of the device from the physical science perception, I have prepared this overview, so one may get a feel for the forces present and how to effect them based on the physical adjustment of the device. There are several flow paths of energy at the physical level including the magnetic field, the isotope line, the diamagnetic field, as well as nuclear torsion or tempic field. Conscious connection is through the diamagnetic field where it may become dominant in either material.

We Observe the Conscious Interactions First

North poles out creates two states:
Bismuth = outflow [From device to operator]
Iron = inflow [From operator into device]

South Poles out creates two states:
Bismuth = inflow [From operator to device]
Iron = outflow [from device to operator]

This is because the Proton and Electron spin opposite directions in the same magnetic field alignment.

Interactions

While the Rain effects may be possible for a dumb device, an understanding of the basics is far better. It allows the operator to make changes and know why.

It is good to gain a first hand feel for the different ways of consciously connecting with the device and a realization that we are all very unique. As a warning it is not a good idea for a new experimenter to set up two or more devices that operate in conflict as they interact with you. An example might be two systems with large iron layer mobius coils, where magnets are reversed on each one setting in the same room. This may manifest a powerful conflict between the mental and emotional state in people in the immediate area and can lead to sickness over time.

Operating multiple devices in one local area takes skill, understanding, and patience. In the above case we are routing Astral energy on the return loop to Source into the physical while at the same time projecting a conscious intention into the mental plane. As it manifests it will hit the outflow loop, and a high energy conflict will be set up in the physical plane. Source flows will be powering it. Indications are that we may not be ready for this just yet as it affects the brain very strongly. If you are not sure keep your devices configured the same for magnet polarities. Or keep the devices at different localities and well separated. If the diamagnetic field, or the magnetic fields share any overlap you may end up with a state of mental confusion and a major headache. You can check the conscious flow charts at some point to gain abilities at this level. The normal reactions for a beginner should be a heightened mental clarity, and increased energy levels. So learn to monitor yourself as a first or prime intention.

Those warnings issued lets proceed.

Energy flow in the system:

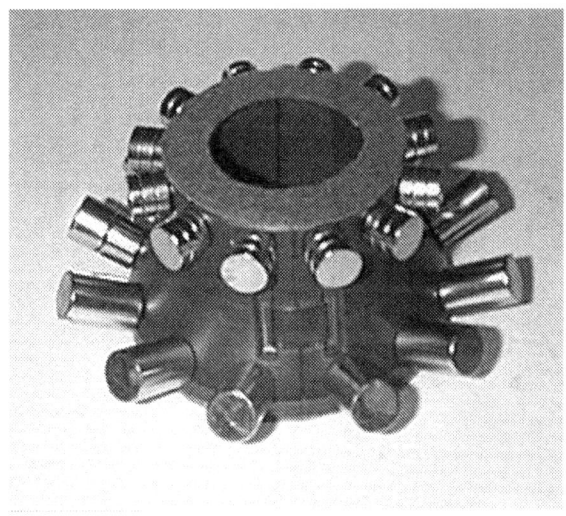

1 - Iron Ring extends magnetic field inwards as one pole interacting with Bismuth coil

2 - Bismuth connects and begins to send nuclear torsion outwards through the magnetic field. Also the bismuths natural diamagnetic field expands slightly.

3 - Scalar coil is energized, or shorted, now on the bismuth coil and begins to deflect the energy of precession motions into the nucleus diamagnetic field. The bismuth is now interactive with our consciousness through the diamagnetic field.

4 - Scalar coil on the ferrite ring deflects energy into the diamagnetic field and reduces the magnetic field extending inwards to the bismuth coil. The iron is now interactive with our consciousness and the bismuth connection energy is reduced in level by this diversion.

5 -Quatrz Crystal has a perfectly balanced dual spin quality and will generally follow the dominant field. A crystal ball however will lean towards a natural inflow state.

[If one is doing healing work I found that an outflow strong enough to move the crystal ball into outflow was very beneficial. This type of healing field is useful for me as a healing Chi source. If the process is allowed to continue for many days or weeks the CU that develops may become like a child and begin to interact. This is a natural result of bringing a large amount of conscious energy into the physical plane.]

Here we see the balances present in the RainMaker 1 from the physical perception. Altering parameters may effect the state of either material to dominate as either an inflow or an outflow.

Set Up Configuration

To experience only the Iron, simply open the bismuth coil, keep the wires separated, and place a Tesla scalar coil or a Mobius coil or Lakhovsky coil on the Ferrite ring, and add as many magnet positions as possible long enough to reach the edges of the coil. The 12 count works good for this as they end up very close together and you may want South poles out for this [Iron outflow].

To experience the Bismuth remove all the coils from the Iron, separate the magnets to maybe a six or eight count, and use the bismuth coil with a function generator. You may want North poles out for this one [Bismuth outflow].

A device may blend both of these at various levels. If it is noticed that suddenly the magnet polarities are working backwards then one has probably shifted to using the other Source flow.

Device Manipulations

In the two layers of material the iron has one free electron and the bismuth has one free proton. These are interacting to connect across the gap through the common magnetic field present and torsion is transferred across the gap. Bismuths magnetic field is connected with its mass and Iron is not but has more power to extend its field.

As we add scalar canceling energy to the iron, less electrons are able to form the energy connection to the bismuth as they begin to cancel at the EM level with one another. The bismuths diamagnetic response to the iron is lowered, and the Source flow in the Electrons is exposed as a diamagnetic field from the Electron shell side. You can see here how the flows are diverted to the diamagnetic field. Our connection with the device through the diamagnetic field shifts to the Electron layer. This is the return conscious control loop back to Source. Manifestations implanted at this layer will be returned to Source and then manifest on the reversal.

If we remove the irons scalar coil and separate the magnets well so they do not cross precession cones in the iron, now we have a strong free iron electron magnetic field extending at full power into the bismuth and interacting more with the bismuth layer. The bismuth scalar coil now deflects this energy into the nucleus and the diamagnetic field becomes much stronger from this layer. The diamagnetic field present is now shifted into the bismuth coil and exposed more from the Proton layer. Since this layer is setting at an alternate density this effect is stronger with respect to torsion. We can operate here directly in the outflow from Source as it hits the Physical and Astral planes.

The RainMaker 1 thus gives access to both flows depending on the balance achieved by the physical setup of the device and which diamagnetic field becomes the strongest at the physical layer and whether it is inflow or outflow. Magnets thus may appear to work in reverse based on which field is dominant as the conscious link. Interaction also will vary from a verbal type communication in the physical to a more telepathic type one as we move between the flows.

I have now witnessed rain effects from either magnet positioning however it is felt that the intention is merely entering the conscious loop at a different point. As the inflow or outflow that become predominant pulls on the physical plane hardest. If inflow becomes dominant in either material over outflow in the other material rain effects may become present as the background Aether begins to be pulled over time. This effect moves directly through the tempic field.

I normally operate with no coils on the iron ferrite and work directly from the nucleus side in the Bismuth. This requires deeper states of meditation generally and a conscious "slide" into the coil. Others work from the iron side with large mobius coils out there, and this can be as simple as a note on the machine that you read verbally to state your intentions in the physical plane. Both Spiritual techniques are effective, and leads to a healthy respect for entering the Conscious loop at any point, however magnets will be reversed in each one and more scalar energy must be present in the correct material to make it dominant.

Lots of parameters here to follow, but some clues as to the different things we are seeing.

Forces and Reach

Irons magnetic field is rooted in the Electron shell, where its mass is very low and its magnetic field has a very long reach. Iron nucleus is magnetically neutral however all its Protons are linked to the Electron shell and balanced there. It has a free Electron to interact externally from the Electron shell level. This is commonly called EM and allows iron to become magnetized coherently as one very large magnetic field that will extend many inches. Its reach will easily move into the bismuth coil at the center to align the Proton fields at that point and control their precession angles setting them into a circle with one pole outwards and one compressing inwards.

Bismuth has two qualities of importance. Its natural diamagnetic field is very weak but does extend external to the material. Thus a scalar coil close to its surface intersects this field that is already present and device motion is not necessary for this field to become interactive as it is with Copper or Aluminum.

Secondly Bismuth is magnetic at the Proton layer where its particle spin is reversed from the Electron layer. The Protons magnetic fields are all very small and barely reach the next atom, however it is coupled to its mass or weight. An isotope line may form inside it to propagate the effects that are generated only very close to its surface. The torsional effects are created at the surface of the material but move all the way through it. The bismuth therefore does not project a field outwards to the iron, it spin couples to the iron magnetic field and transfers its torsion of mass momentum through this field outwards. It also propagates this field as an isotope line into the rest of the Bismuth atoms. This is a torsion field of tempic effect and what we sense with our hands as heat or cold or vibrations from the device. These torsion field have been associated with pyramids as well as large spinning flywheels.

Further experiment may more clearly verify this as we progress, but it would appear that devices can be designed to enter either side of the conscious flow as it moves from or away from Source. Physical properties of dominant field are however device dependent, so machines should be monitored when they are active.

These appear to be the steering capabilities for the conscious flows in the device.

Map of the Conscious Flows

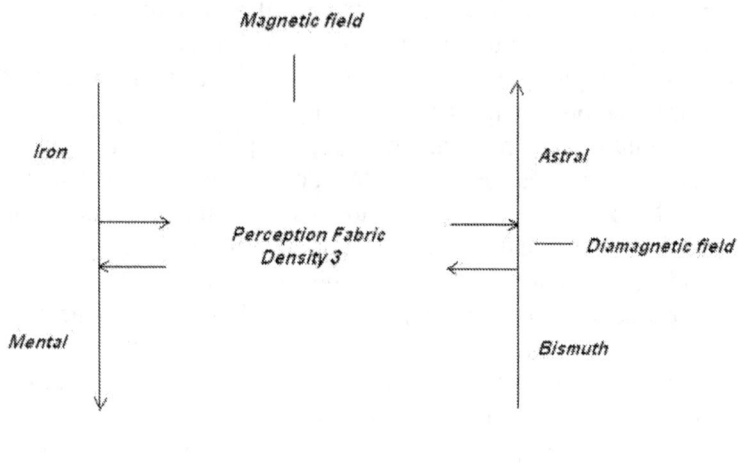

From the chart it is good to realize that inflow and outflow are terms relative to where your awareness is setting. In our normal waking state we are setting at the center of this chart. Astral and Mental planes are projecting the sphere of physical space, and we are projecting our consciousness into our physical body where we think we are setting. The flows are present in all matter, but in the Iron and Bismuth they can be hooked magnetically and manipulated.

Definitions

Diamagnetic field: Offers a repelling force to either end of a magnet and appears naturally in bismuth. Offers a path for consciousness to enter the magnetic field control vectors [healing arts] and effect reality using intention.

Outflow / Inflow: A dynamic of the conscious energy to flow or establish flows into or out of conscious beings that exist

184

separately from Source. These flows have an intelligence of a higher order, but also link us with the emotional and mental planes.

Source: "Source concept" the God force that powers atoms where we find the rest state of an atom is light speed, its rotational velocity. The control vector is the magnetic field and appears in two directions where spin is observed to be reversed. Proton and Electron, and merged in Neutron which contains dual spin.

Tempic: Coined by Wilbert Smith describes the light speed spin or motion that is fairly constant here in density 3 matter. When altered the light speed constant shifts slightly and sensation of time flow rate and gravity are effected. The tempic field is rooted in the Aether of the conscious planes and connects us all.

Disclaimer:

The devices are experimental and rely on the operators level of conscious interaction to manifest effects. These writings are based on our observations and perceptions and not to be considered "claims" or "prescriptions" for healing of for over unity devices. They are pointed at understanding our conscious level as mankind, and reaching for a greater unity through mental plane awareness. It is my personal goal to reach a state of "comprehension" while experimenting with these devices, as should be any who embark on the path. Only from this level can they be utilized to effectiveness.

Many thanks to all the participants on the c_s_s_p group for there input.

1 - 21 - 2007

Kosol Ouch, Koeun Noun Ouch, David Lowrance, Martin Pott,
Jerry Evans II and Vince Panella

Magnetic Vortex Generator - Construction Detail

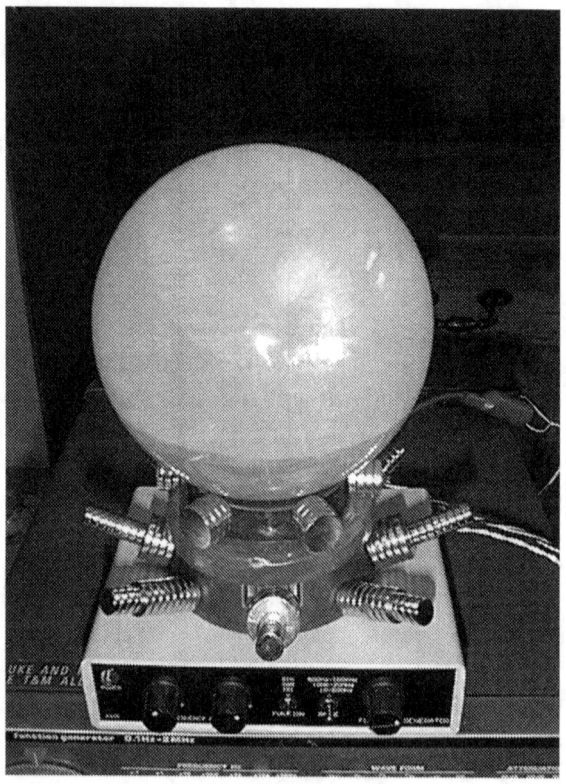

Aluminum Bismuth Scalar Coil

This coil is the heart of configuring the device for a Proton magnetic outflow or inflow.

There is much leeway in the coils, they can be custom sized to fit the device however try to keep the diameter under 1". 3/4" is optimal and bigger is not better. If a crystal ball will be used keep the coil shorter then the ferrite ring so the ball will sit on top.

Materials

Aluminum tube - 3/4" diameter cut to 1 ½"
Bismuth shot BB's [enough to fill the tube about twice there will be some waste]
24 gauge copper or tinned copper hookup wire about 20'
Electrical tape grey or red as black is rather ugly
1 Pkg modeling clay
1 small cast iron melting pot
Aluminum foil
Cookie sheet or other small pan
1 Propane torch - [or stove top can be used burner on high 520 deg F]

1 hot pad mitten for pouring
Safety glasses

Procedure

The coil is made from an Aluminum tube 3/4" outside diameter 1 1/2" long. It is filled with Bismuth. Using a small pan cover the bottom with Aluminum foil. Set the tube upright and then seal it on the outside with a thick ring of modeling clay to contain the bismuth from escaping as it cools. The Bismuth is shot BB's that can be found at a reloading supply store. It is melted in a small cast iron melting pan that can be found at a kitchen supply store. It takes about 520 degrees F to melt the Bismuth, I used a propane torch. Bismuth expands as it cools so do not fill the tube completely. Take your time in the melting and slowly swirl the pan to get all the shot melted. There will be some scum at the top but as you pour the liquid bismuth the scum will be the last thing out so do not worry about it. Let the core cool well before touching it! The best coils will crystallize along the top of the coil as they cool and heating the metal longer and hotter will help this.

Next Wind the Coil

Coil winding Detail

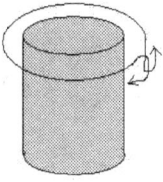

Start in the middle of the 20 foot wire. Loop the core at one end. Twist the wires around each other once so that wires trace back the way they came.

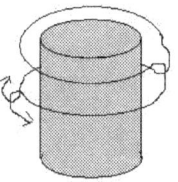

Repeat this on the opposite side each time down the coil keeping windings tight and as close together as possible.

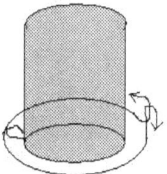

After the first layer is completed wrap the coil with one layer of electrical tape and then wrap another layer carefully up the coil.

This time though place the twists at 90 degrees to the underlying layer.

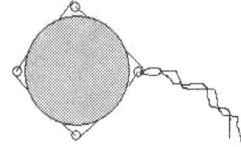

The finished coil should look like this now from the top. Carefully wind the two leads about 1 turn per 1/4 inch and cut at about 2'. A layer of electrical tape over the outer layer will keep the wires in place.

Ferrite Ring

The ferrite ring can be scavenged from an older computer monitor, the bigger the better. It is a slow job and care must be taken not to break the ferrite as the chalking is chipped off near the bottom. The ferrite ring sits inside the yoke on the back of the picture tube. If you do not know anything about TV sets and discharging the lethal voltages get someone who knows as the voltages present even in a powered down set can be lethal. Be sure to keep the metal clips that hold the two halves of the ferrite ring together. If a ferrite ring is not available a 2 1/4" or 2 1/2" diameter iron pipe nipple can be used about 2" long.

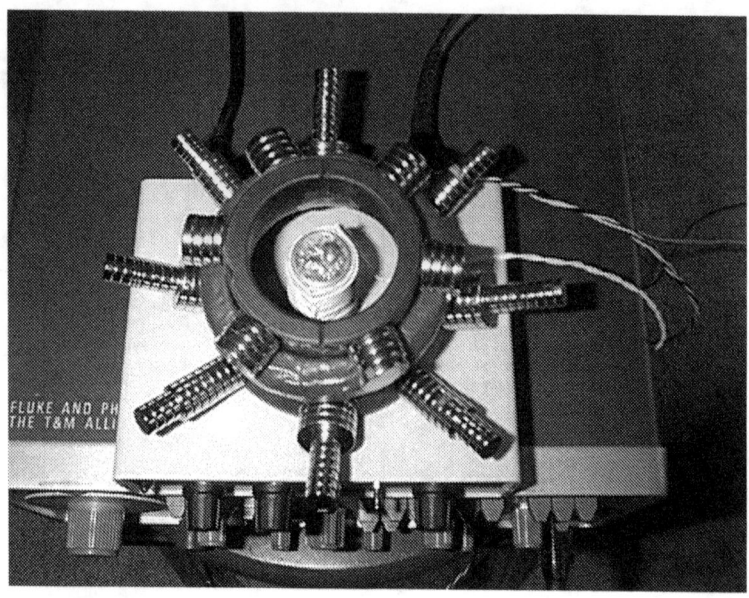

Magnets

Many patterns can be used to set up the magnetic field however Neo magnets are recommended of sufficient quantities for a strong outflow to be achieved.

Here is my latest recommendation:

24 each type DC2 3/4" x 1/8" disc [placed around the bottom in six locations of 4 each stacked]

24 each type DA2 5/8" x 1/8" disc [Placed around the top in six locations of 4 each stacked]

Optional magnets can be added to these of a smaller 1/2" x 1/8" size to expand the fields width. D82 is a good choice.

If a large crystal ball is used this will expand the field envelope. A good magnet layout may cost around $60.

Use a compass to make sure magnets are placed correctly. North pole should be out on all the magnets and no reversals should appear anywhere along the ferrite ring. Depending the size of the ferrite ring it could take more or less magnet stacks. There will be a maximum limit of magnets before they start to push one another off.

Ferrite Coil

Function Generator Coil

The Red coil around the ferrite ring shown above is optional and offers a scalar interaction with the iron atoms. It is used to create an Electron inflow or outflow vibration using a function generator to excite the field. It is made by laying red electrical tape turned sticky side out along the ferrite then wrapping a twin lead 24 gauge wire around it between the magnets as thick as desired then wrapping the tape over to hold the shape of the ferrite so the coil can be removed easily. To configure this coil for scalar operation the wires from each end of the coil are fed opposite directions for a magnetic canceling effect. As an option one end can be shorted and taped then the coil is fed from the other end.

Flat Pancake Coil

The green Tesla type pancake coil has now been shown to be effective and useful as well for an Electron activation element in the device if one is researching power generation effects or just wants to feel the Electron heat energy. It easily throws out a 6 foot field around the unit with no bismuth coil inside just from the magnet, iron, coil interaction so lends to creating an Electron vortex device. This field is comfortable to work around and has no ill effects I have found as with the Lakvolsky and Mobius coils that produce chaotic energy. The crystal, if used, may be turned so that the lattice forms a V to the center of the unit rather then flat because this coil produces a conical field.

About 24 feet of #12 solid copper wire
Electrical Tape
SPDT switch [optional]
8 feet hookup wire [optional]

The coil is #12 solid copper wire with insulation, two wires stacked one on top the other. I placed 9 turns on this coil but the winding could extend fully to the ends of the longest magnets for the greatest effects. The coil is wound from the center of the wire, If you want an on off switch cut the two wires apart and solder them to smaller jumpers for extending to a switch, if not, just bend it over at the center lay it flat on the ferrite ring and start the wrapping process. Tape the next layer at each quarter turn and keep both stacked very neatly.

When you get to the outer end of the coil strip 1/4 inch and tightly tape the ends together so they are touching, or solder to smaller wires for on off switching. Both ends of the coil must be switched on and off using a DPST switch other wise either end will act to cancel magnetic fields.

Radiation pattern is a conical shape off the precession angles of the magnetic poles of the iron atoms and the Neo magnets and is found to be quite hot.

Function Generator

A program is available on the Kosol Core Tech group that has been especially designed to drive the coils using a computer sound output.

If this is used an 8 ohm to 100 ohm resister must be placed in series with the coil to protect the sound card of the PC.

Other wise a function generator can be purchased to drive the coil. Ideally one can experiment with frequencies, however NMR resonance effects happen between 1 and 20 Mhz for the metals present. Frequencies even as low a 8 Hz have been used, but I recommend the higher frequencies. As one works with the frequencies they normally tend to increase them over time.

Crystal Ball

The crystal ball pictured is a 125 mm Diameter Calcite sphere.

Quartz crystals have a natural electric vibration and any spheres can be used.

Dave L
Kosol Core Tech 9 - 29 – 6

193

Australian Weather Research

The following is the experimental record complied by David Lowrance at c_s_s_p group, working with Smokey in Australia, who has been monitoring the weather for many years and has a genuine love for the country and people. Smokey has further developed and now tried operating a couple of the conscious devices. The tube device, which he first built using the principles of diamagnetic fields, and the RainMaker 1 unit from our previous work. He is very attuned to the weather patterns and has been for some time, thus has a spiritual connection with it. As Australia has been in drought now for many years in these areas, he gave permission to make this information public outside our group and extend a warm thank you to Kosol and those working on these Spiritual/Science Technology devices.

We are beginning to learn, these devices only work with the operator present and are an extension of our own abilities to tap into the diamagnetic and conscious fields. They do not work for all people. This is the method we have observed working for those like Hamel, Searl, Sweet, Hutchison, and many others where often, only the creator of the device can make it function due to their conscious connection with the devices.

As per the "Organizational Unity Concept" offered by the advanced cultures, I also feel the need to share this information freely with the sum total of mankind, as the higher purpose of these discoveries, as they are being made.

11 -11 - 6

Perception Event Record
[c_s_s_p group]

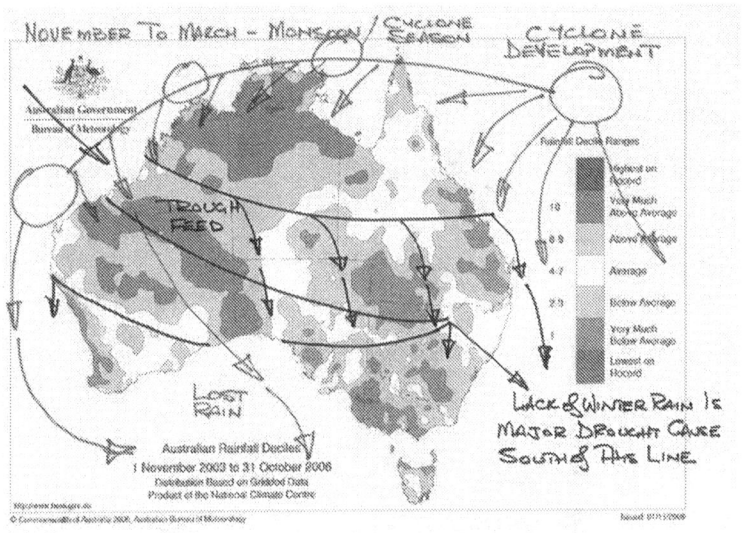

Smokey provided us with this chart to indicate his work on why Australia has experienced such drought over the past several years. He indicated that there is presently lots of rain falling over the oceans but it has stopped moving inwards over the land. We had another individual come on the Kosol site expressing a need for rain in the Victoria area just inside the bottom tip along the coast. During the second development section this caused Smokey to take another look but he decided to set his "target" for the center of the worst area with the intention that the rain would hit all the points in need even 220 Km South of the target.

Smokey first offered two emails for public release containing the following photos and information showing the depth of his involvement:

This is a data record of the OZ Rain Event from 31/10/2006 to the 07/11/2006 - The Event has not ended and is still continuing but this is the start. I have only just woken up to the

fact that I am using Compass readings and then transferring to
Maps/Charts and was concerned that I had about a 10degree error.
Well the error here is +11 degrees and that now brings all
perceptions/observations back into line which immediately gives
me greater confidence in pointing ability.

Photos follow in order:

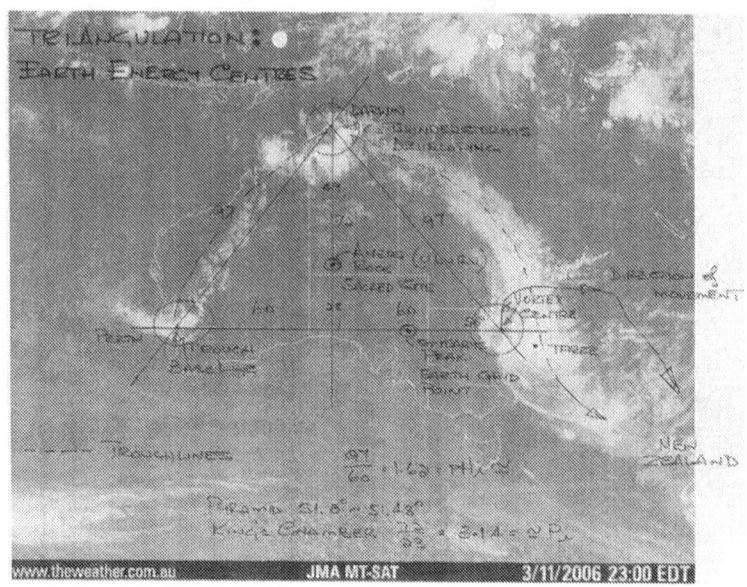

1. Satellite - SemiCircle of Cloud over Australia. Have noted
the Triangulation of Storm activity on this Map and this is subject
to Physical analysis. I have mentioned somewhere the names of
Bruce L. Cathie and David Hatcher Childress and was wondering
if they have stumbled upon Earth Weather Pressure/Balance
points because I believe they do exist. This phenomena I will
continue to monitor as it may lead us to an Earth solution to
Drought.

Further analysis:

Did some calculations and found:

1. Base Angles are 50 degrees - Pyramid is 54.609 deg.

2. Ayers Rock (Sacred Site) is sitting exactly at the King's
Chamber position - 72/23 = 3.13 nearly Pi, 3.14. (72 is the full

distance, peak to baseline in mm and the 23 is the distance up from the Baseline and equals Pi in the Pyramid). I don't believe I have taken any liberty with these measurements but have guessed centers of activity and that's what is appearing here.

As a consequence, I am now looking at the Energy emanating from these Devices as being 'different' and possibly in tune with the Planet. Why wouldn't it be because what we are connecting to is Mother Earth - Natural Energies.

Can this possibly be the 'Tool' we need to 'Heal The Planet'???? Is this possibly what the Guardians are saying to Kosol, that we must develop these Devices, using our knowledge and abilities and this is the result - This is what we are meant to discover?

Why have we detoured away from OU/AG - we haven't - we have been pulled away to Rainmaking to show us what is needed to Heal The Planet'. We are being Guided.

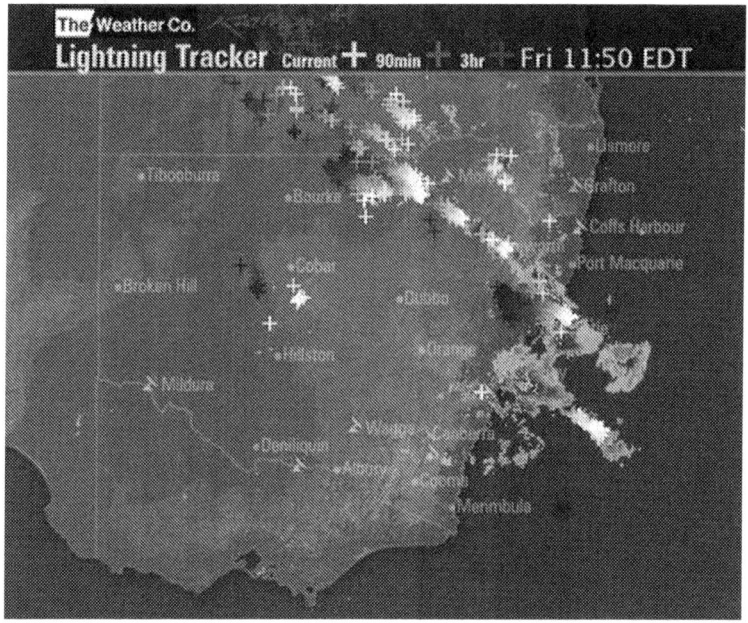

2. Radar - Storm Line along 295 degree 'OZIIC' pointing angle.

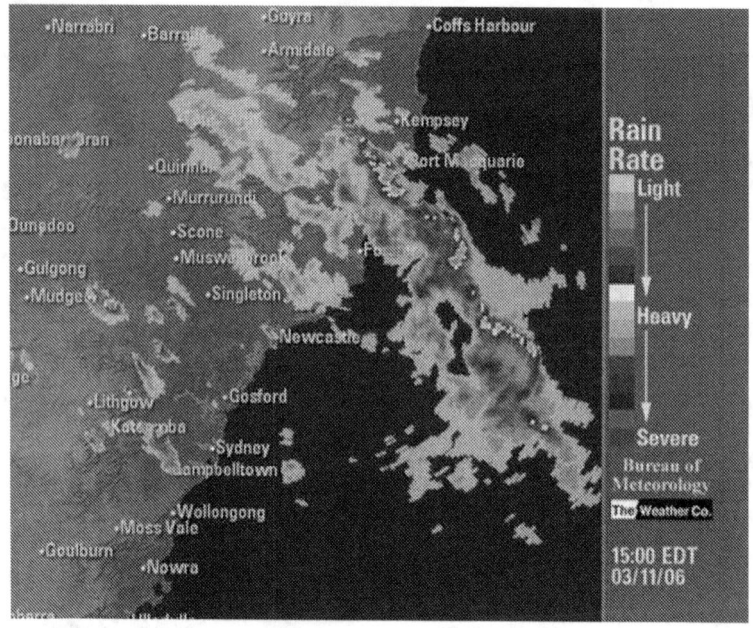

3. Radar - Rain Band Semi-Circles after passing Operations (OPS) Area at Taree - NW going to SE.

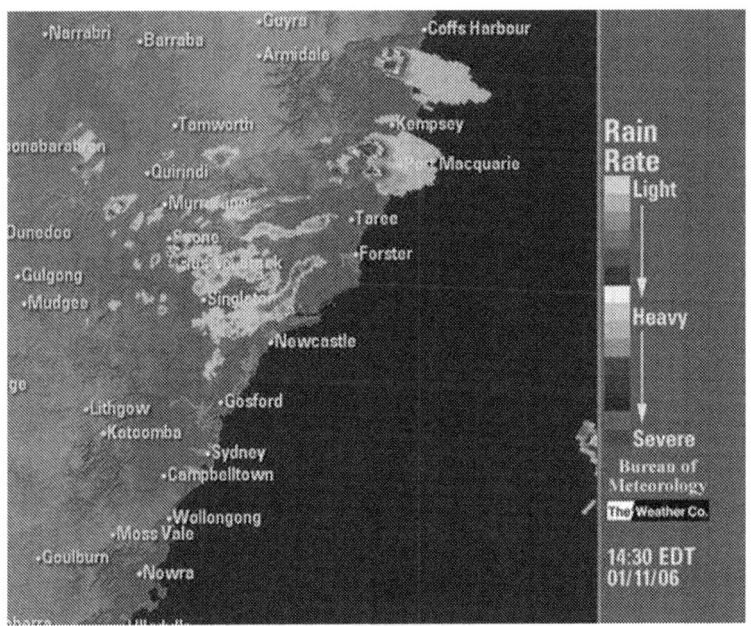

4. Radar - Rain Bands from SW 'Seeking Source'?

5. Satellite - Clouds falling in line behind 'OZIIC' after its pointing was changed from 295 to 255 degrees to attempt Rain push into needed areas. Not successful as S/SW winds were too strong and you can see front SW edge being swept away to East. This was active for 12 Hours - Formed in 1 Hour and Decayed in 1 Hour, after pointing changes - Amazing! You can see the signature of 'OZIIC' in the distinct Cloud Bands behind the unit on either side, like huge rolls.

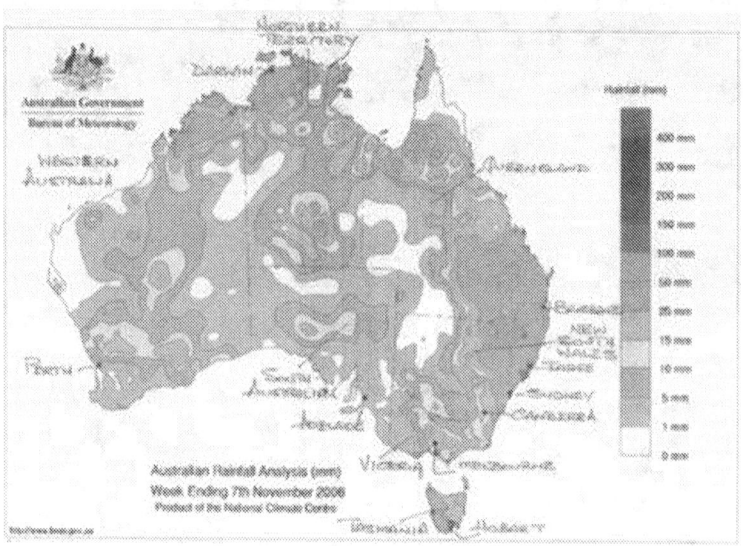

6: Chart - This is the 7 Day Rain Chart. Areas in the 'Top End' of Western Australia (WA) and Northern Territory (NT), top left, are the results of the Triangulation, a Monsoon that is one month early and already Record Rains. Bottom Left is Perth where some Rain fell over Agricultural areas but requires input. The East Coast is showing a huge impact for both New South Wales (NSW) and Queensland (Q) with some Record Rainfall. Rain falling on the Great Divide which begins from the South Australia (SA)/Victoria (Vic) border and sweeps in an arc right to the tip of Cape York in Queensland. Rain here flows both West and East and with 5" up here, the Farmers now have flowing

Rivers and Irrigation waters. A Start. The Experiment pointing is now attempting to fill in the gaps - basically from Perth to Sydney - all Agricultural/Irrigation.

7. Satellite - First indication of Vortex forming.

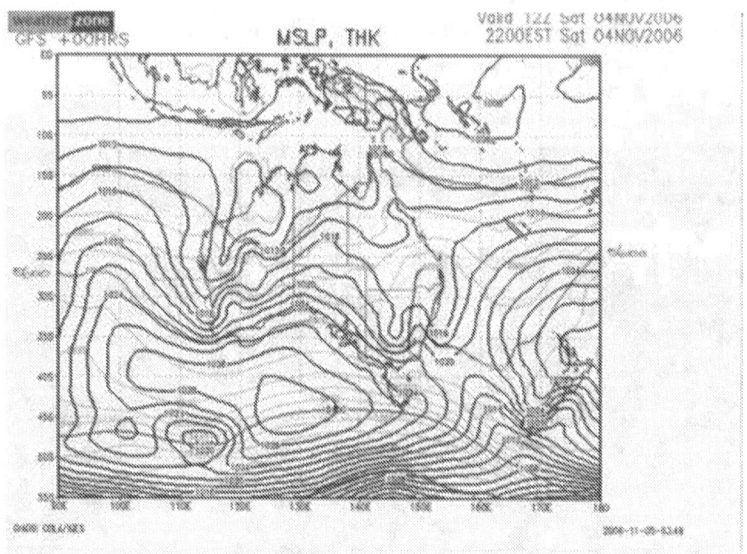

8. Chart - Believe this is what we are creating to encourage Rainfall. Cool Pool air from the South in an Upper Level Trough, is perceived to have been drawn up by the Devices into warmer climates where Storms are created, with eventual Rainfall. I have drawn the 5600 Upper level in red, indicating the 'pulling' at the Upper Level, right up the East Coast. This is an MSLP (Mean Surface Level Pressure) Chart but shows both Surface (Dark) and 500 hPa levels (Light).

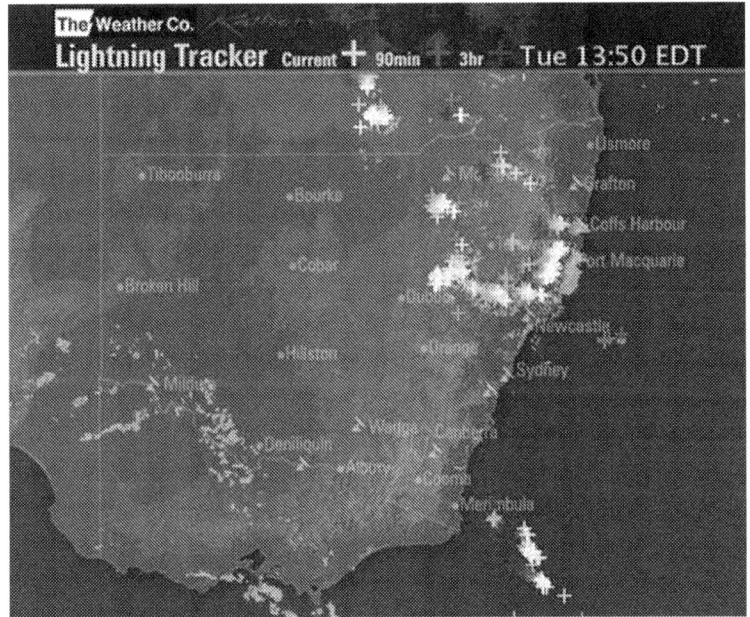

9. Radar - Storm Circles - Amazing! - First time witnessed. The Devices are creating a Vortex, but different this time, may be at a different Upper Level. The Long Line Trough that is formed by these Devices is being formed into a Circle to the NW of the OPS Site where the Storms are forming. Trough Line in South, then Vortex, then back to Trough Line in North. Two Types of Vortex form and these are also subject to further analysis.

Devices Being Tested

Also for the Record need to clarify the two Units used for this Exercise.

1. 'Rainmaker1' - David's but modified with two Lakhovsky Coils, two added Ferrites, loaded with Bismuth Pellets in the Core and Magnets on the outside and a large Quartz Crystal sitting on top of the 2nd Lakhovsky Coil - promoting 'Calm'.

2. 'OZIIC' - As pictured in the original design, but with second Lakhovsky Coil alongside first.

Both Units are programmed with Intent and Conscious Palming on a regular basis. I am beginning to feel that the Units are responding to my presence.

Second Major Development
11-08-06

At this point Smokey had managed to bring a consistent rain perception to his own area. I must admit at the time thinking ok this could have been a fluke but of course we were happy for him.

I did a "slide" on this date to observe Australia from above it using my RainMaker here in Alaska and just check in on Smokey. I set up for inflow mode as it seems to offer a greater depth of perception. I observed that Smokey had lit up the whole East coast of Australia with Light energy. I then moved to the center of the continent and began to perceive the convergence point on the earth to discover that it was feeding Smokey's device. He had already sensed that a higher consciousness had been guiding us and he had mapped these sacred points on his continent. I was then drawn outwards by guidance and shown the whole earth. The connection to the Earths consciousness brought a new perception. The Earth is moving through the "convergence" and we are only the beings on its surface setting in its' field. I confirmed Smokey's center point in Australia. Then took some time to try and understand what had just happened while in the "conscious slide" and if it was real.

Dell brought to my attention that I was setting on one of the earths central energy points myself, and a scan of my local area revealed that Mount Arrowhead approx 7 miles away from me here in Sitka would seem to be a point of the "convergence" as well. I connected into the earth at this point and began to channel a lot of information that would tend to support many things we have already been told by our Spiritual community. "Electric systems will fail to operate after the convergence" was the most disturbing, but that magnetic devices would continue to work and far easier then they do now. Induction will no longer hurl electrons from the diamagnetic field, and this is the basis for all our current electronics, due to a coherent diamagnetic field becoming present much like a UFO flying over.

Smokey was still concerned about getting the rain to the drought areas but had also noted that his own garden was up around 4 feet and covering the bird feeder, where last year 4" was all he had gotten.

I suggested using the earth points and setting up two tubes. One tube was already pulling in much life force to Smokey's locality, now the goal would be to extend or channel this energy to a second location by placing the tubes so that they would create a conscious or diamagnetic energy link from the earth's energy point to the target.

He set up a second tube, one to receive the earths energy point and another to project it to the South West where the drought was worst. Energy is sensed hopping between the tubes and Smokey indicated it would probably take several days to see the result.

11 - 11 - 06
Smokey writes:
Remember me talking of 'Blooming' being evident when the Clouds arrived at sea. Dell and myself, both talking of rain not wanting to come onto Land. We are being sent a very clear message that Mother Earth is not happy. We perhaps know that, most don't. Have just woken up to what I believe is the problem. This is Mother Earth telling us that there is too much 'crap' going on, on the Land and that's where Humans are! She is happy at sea with the natural creatures and Energy there.

We are now seeing 'Blooming' wherever our Devices are present, particularly on land. The Land simply lacked vibrant Energy to support Rainfall before these Devices. People, when they go to Ayer's Rock are uplifted by this mammoth sitting out in the middle of nowhere - really they do not know what it is that is doing the 'uplifting ' bit, it's not awe, it's Spiritual!

The People are 'dead', lacking in Spiritual Development, not being a part of God. Current 'Religion' does not teach nor support, true 'Spiritual Development'.

This is the beginning of the cleansing of the Land. We probably knew there was going to be changes but what? Is this the start of something better now?

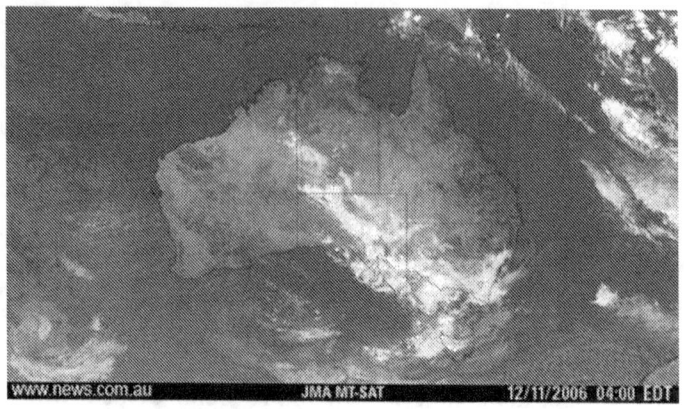

Both Yarram and Deniliquin have had Rain, only light but more to come, Thunderstorms. I was looking at the 'Blooming' over the target area, (OH so Beautiful!), when the Lamp lit. Bit Slow! Still in disbelief here. South Australia only light rain, had bad wind earlier but may get more rain later. Hoping Cloud mass will slow and build in size. Rain Bands are turning to rough 240 degree angle as they enter Victoria.

Smokey

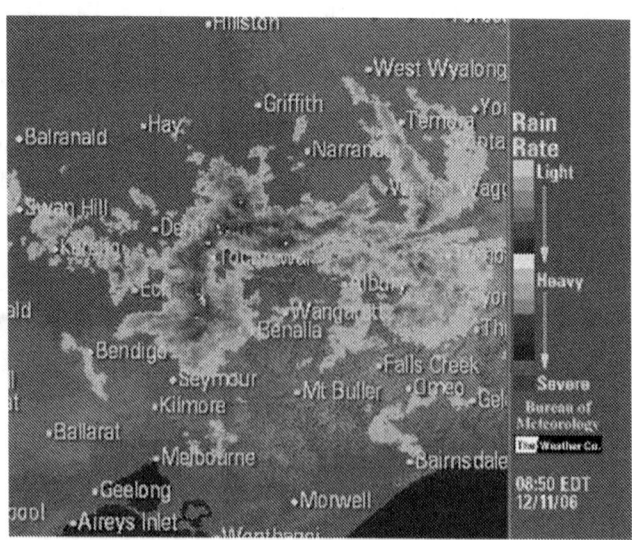

**There were both tears and cheers at this point at the
c_s_s_p group!**

You can see from the above charts snapped on 11/12/06 that
Smokey has managed to bring clouds and rain into an area some
distance away from his position over on the East coast region. We
recognized this to be one of the greatest accomplishments we
have experienced to date. He had reported that this might take
several days to happen, and indeed he is getting good at knowing
how his technique operates. Approximately 4 days for rain to be
"perceived" at Smokey's target area. [Note the date format in
Australia reverses the day and month from the normal US method
and the photo dates look reversed to us day/month/year] Also
there is a time difference and it would seem that often we are on a
different day. The 12th in Australia is the 11th here in Sitka for
some hours of overlap.

Dave L

Later on the 12th Smokey writes:

Hello All,

Slept well last night. Believe I was being told, not to worry,
all is in hand. 'Conscious Energy' at work! I sit here in abstract
awe, most Humbled, watching these Rain events unfold before
my very eyes.

South Australia, 1/2" to 1 1/2" - It does not happen this way
in South OZ - it always just gets teased with Rain, nothing like
this and more to come. My home State.

Deniliquin, 1/4", a start and highest for the target area.
Yarram is receiving and also NE Tasmania, also Drought area.
Rain Band is curving after target and moving to Source, here. Did
not expect this result, it is immediate! It IS where we need it the
most! I am a perfectionist, hard to please, I don't believe this! The
concept of 'Conduit' is valid - a kind of Spiritual Awakening for
the Land, a 'Conscious Awakening' for us.

The significant Weather event I mentioned last night follows:
There is a large circulating LOW just below the South Coast
of Western Australia. These are the guys that suck the clouds and

rain down into the Southern Ocean - lost rain. If you have a look
at the animations at:

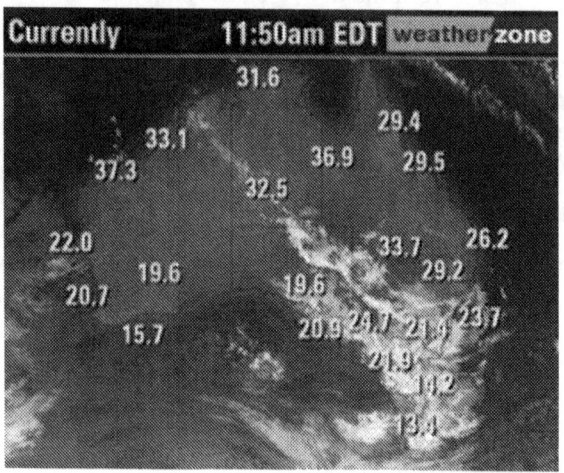

You will see it is there, still rotating but cloudless! All the
clouds are still on the Land with the rain! This simply does not
happen! So what IS happening? A more appealing Energy Source
attracting the Clouds and Rain? The Inner Earth Energy from
Ayer's Rock? Two Earth Energies - Surface and Inner? One from
the Heart, solid, true - the other is the Surface, teasing, the
'Weather', unpredictable, inconsistent, variable, unsure - where
we normally are.

Smokey

Compare the above chart to the first chart at the top of the
page, is this just amazing?

11 - 13 - 06

Smokey has a lot to report today and we are referring to the
tube layout as a conduit for the energy:

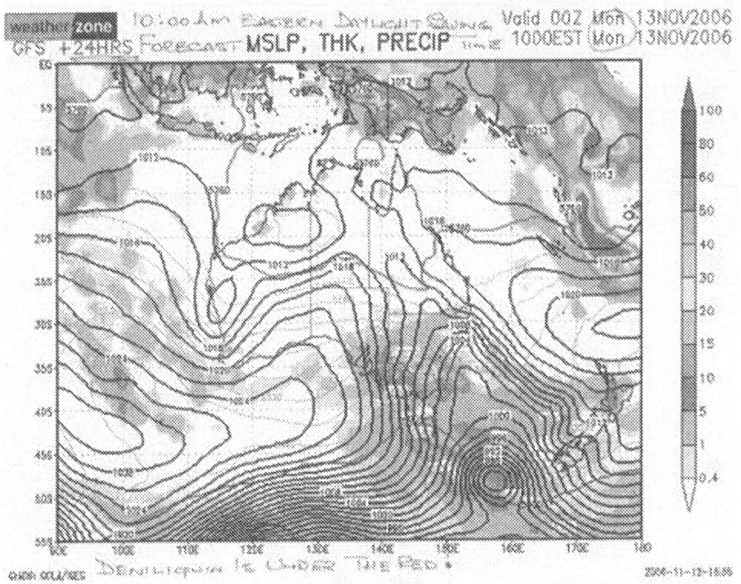

Hello All,
Have added photos:

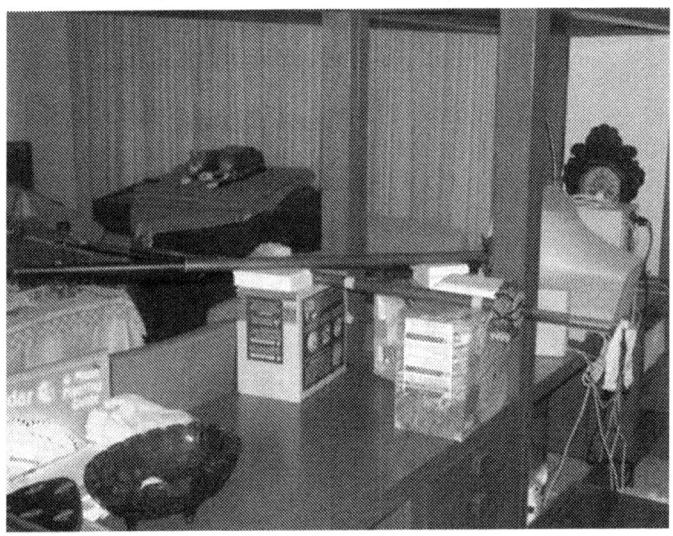

1. 2 of the Tubes in the Conduit configuration - The 'Cold Power' boxes they are resting on is purely coincidental - That is 'Stimpy', in the background, totally oblivious to anything as usual. Likes to be near People and this is bed No.36! Does not sense what we do.

2. 'Love' in Cloud writing and some articles from the Rural Magazine, 'The Land' with regard to the previous Rain Event on the North Coast.

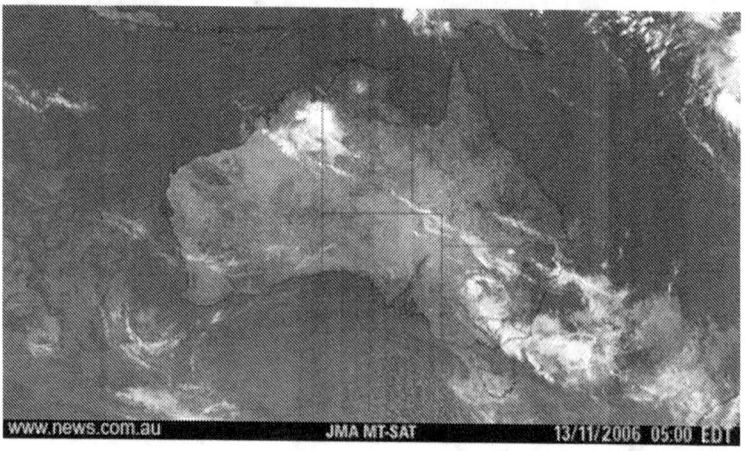

Since completing my daily Weather dump, the situation here is not as disappointing as first envisaged - Good rain is being acknowledged in South Australia - 10 to 22mm over most Rural areas, best rain in 4 to 8 months, 25mm in one area, best in 12 months. No talk of crop losses. Tomorrows totals in NSW and Vic should be similar and just as useful. Not now so sure about the Gales up North but the South Westerly change will probably go through and will not allow a stable rain environment. We wait and see. At this time there are two streams of Lightning Storms both directing themselves at the target area, one in the middle of the Country.

Forgot to advise that about 2 hours after setting up the 'Conduit' Experiment, I laid, touching Copper, a large Quartz twin Crystal against the Receiving Tube. This for 'Calm' again.
Smokey

More observations may be added as comprehension grows.
Smokey indicated already planning some modifications to spread the energy out more.

Notes

While I had merely observed this rain phenomena as a curious effect of the devices early on and nick named the first one RainMaker 1, Smokey and Dell had realized its' true potential and the Earth may benefit, as now some more of us are realizing she is Conscious. I realized that some process would be involved in us "taking our reality" from my first slide but could never have expected our involvement with the Earth consciousness as being a necessary step. I have received an "initiation" or "attunement" into Earth Consciousness during this process that Smokey and Dell connected me with.

We have been locating other points on the earth that would seem to be involved in this "convergence" process. Three of us seem to be sensitive to these at present that I am aware. A couple in South America, Nebraska, Hawaii, and one in Africa thus far. As I scan them all my devices light up with energy as though the earth is telling me and confirming where they are when I am connected into the mountain.

The information here in is considered Public Domain and may be used for any source material needing or attempting to preserve this knowledge and aid the comprehension of mankind.

David G Dawson of Australia
David Lowrance of Alaska
Dell Coleman of Canada
Kosol Ouch

Some satellite-views of hurricane Kyrill 18.01.2007, the red spot is the RM's location.

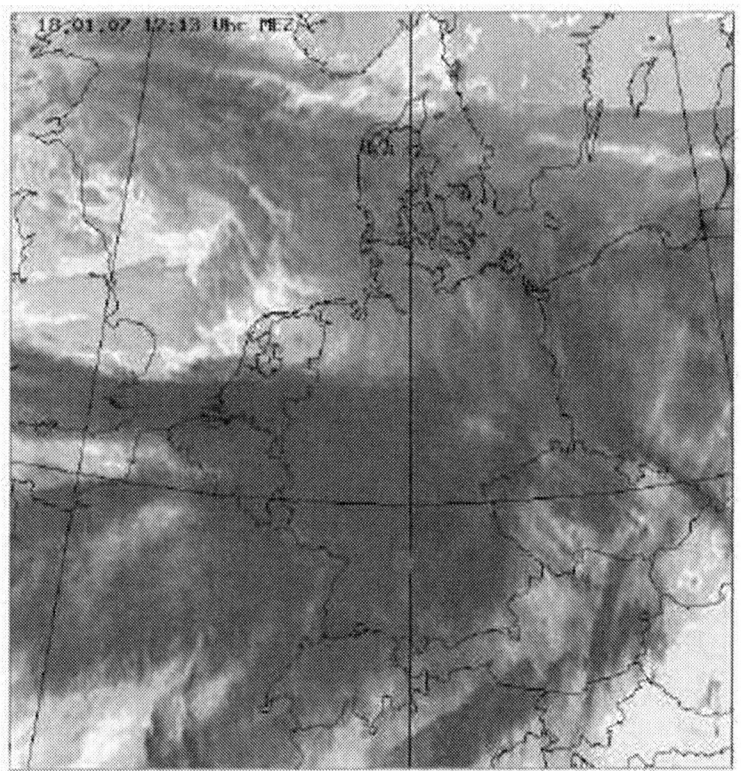

12:13 PM almost nothing happening, no rain, no strong wind, but real dark skies

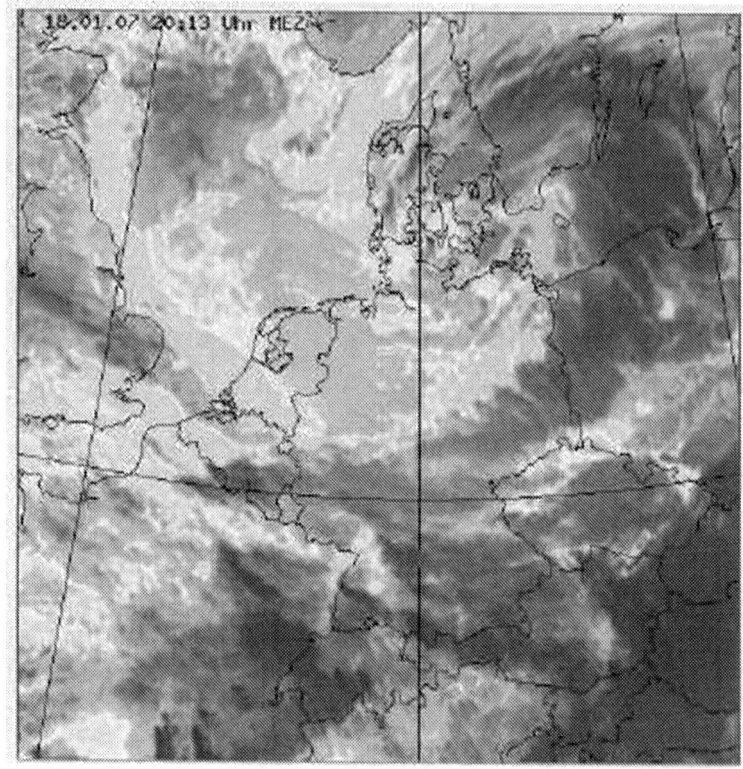

20:13 PM stronger single winds, RM active since 19:00 PM

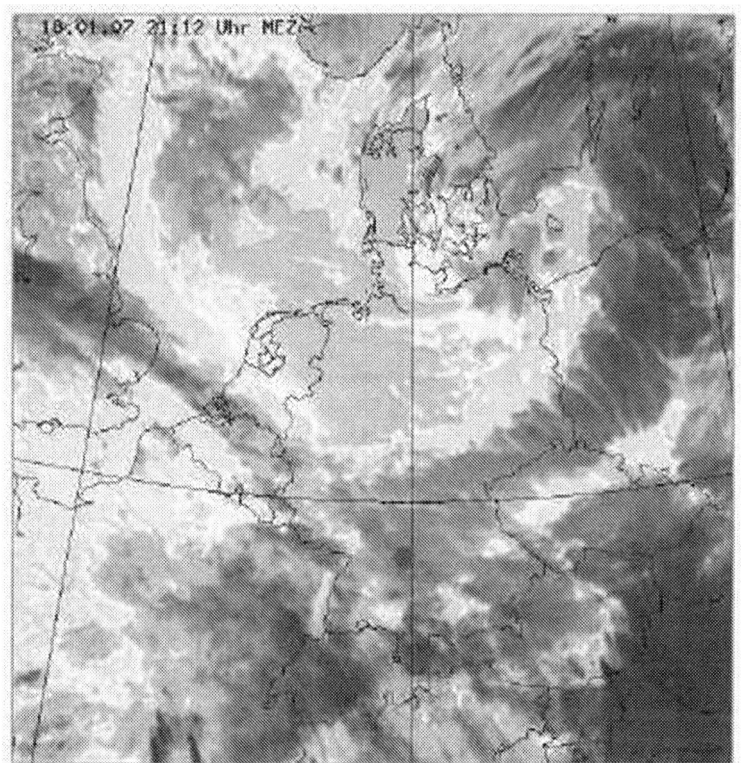

21:12 PM - lighter gray around red spot, other regions keep darker

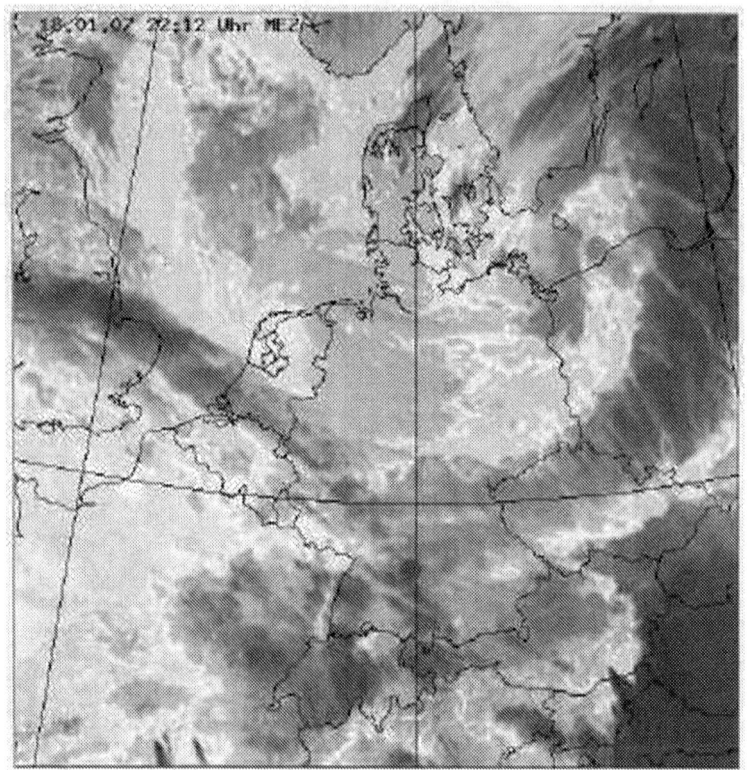

22:12 PM strong line coming from upper left, directing red
point, strong winds are predicted but not coming

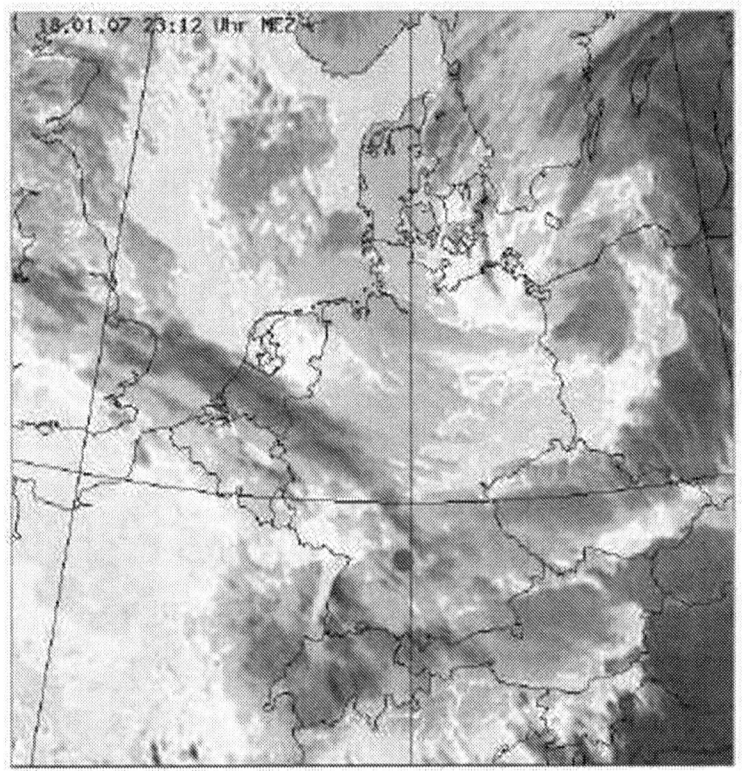

23:12 PM only small rain, wrong prediction of forecast

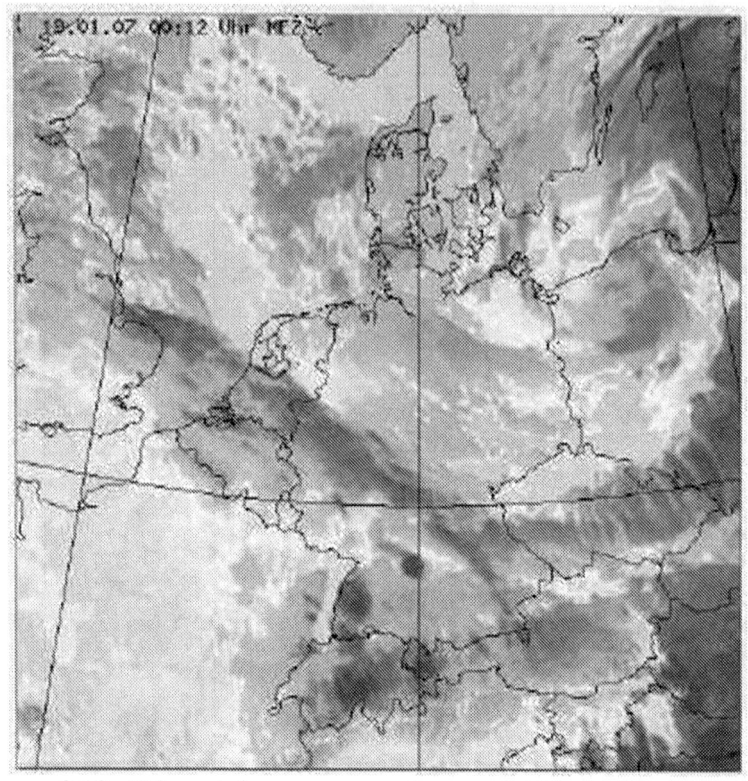

0:12 AM next day, see green free spot, blowing by wind to right from red spot

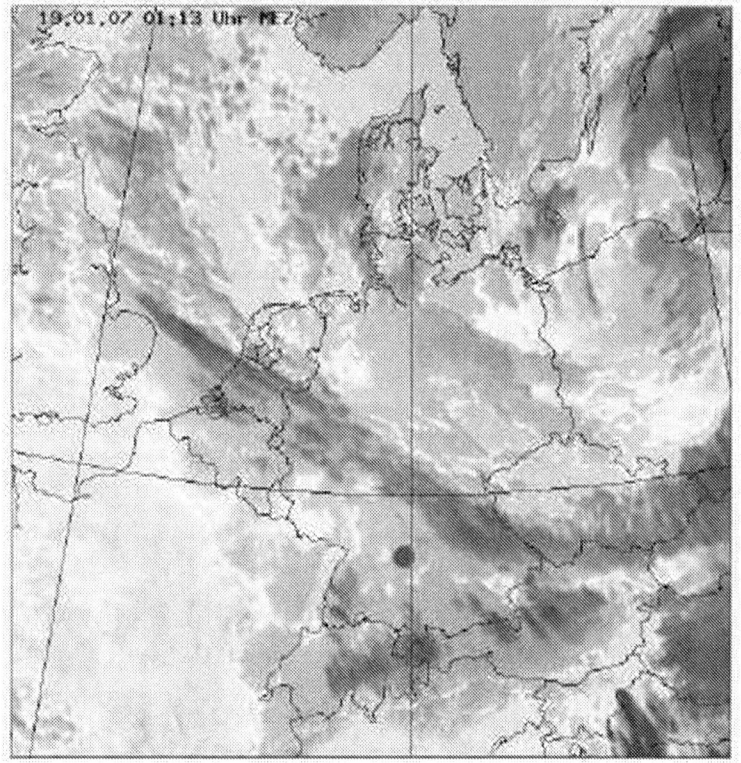

1:13 AM

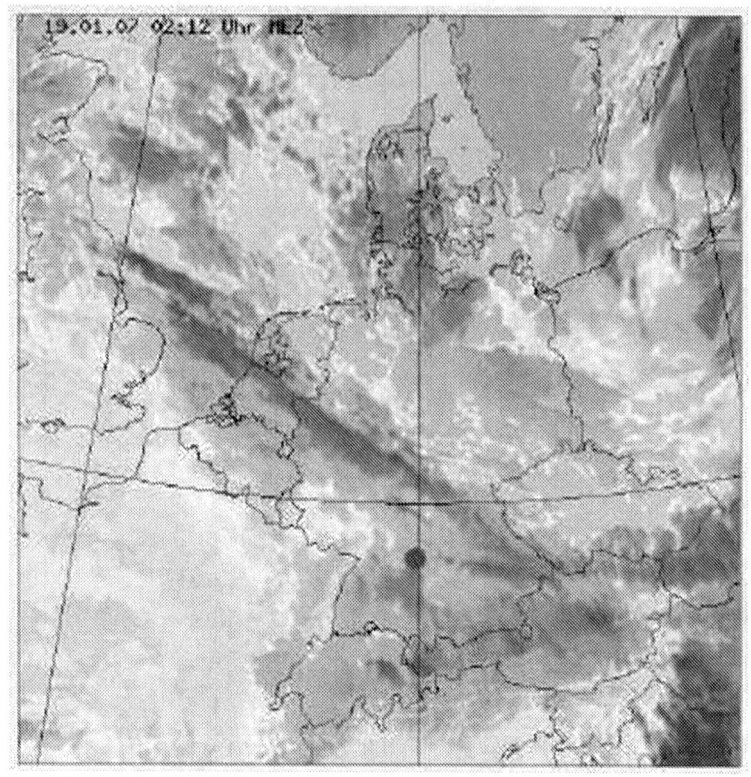

2:12 AM Kyrill lost power

The Consciousness Ship

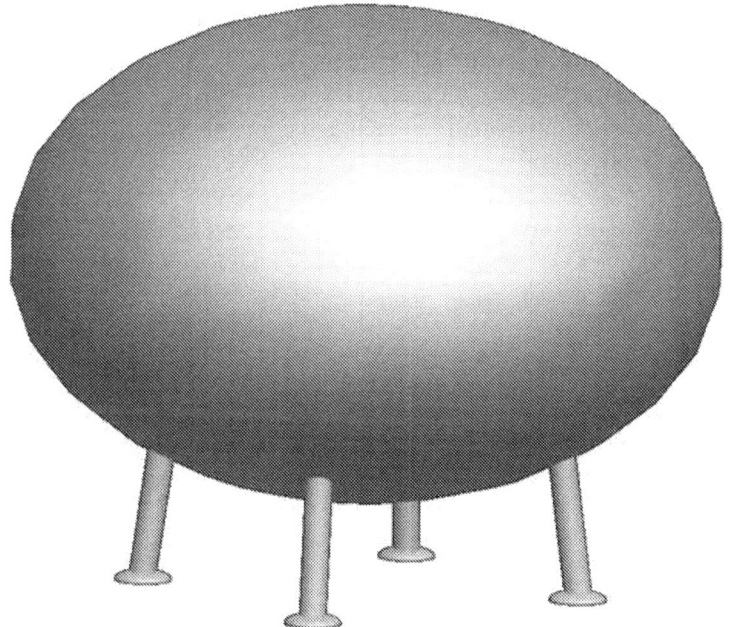

The consciousness ship consists of this material:

1. The two aluminum casing or the two aluminum shelf sandwiching bismuth.
2. The core of the consciousness drive ship consists of the any quartz crystals sphere.
3. Surrounding that is the isoca-dedecahedrons neo dymanium or ferrite magnetic array.

4.   Encasing that is the iron flowering petal hemisphere on bottom or vice versa and the copper encasing flowering petal hemisphere on top or vice versa.

5.   The scalar bismuth coils, there are four of them primary but more can be added surrounding the copper and iron hemisphere that surround the quartz crystals and the magnetic isoca-dedecahedron array.

6.   At the top and bottom of the core is the aluminum bismuth and aluminum obelisk. which one can be wrapped with scalar coils while the other can be wrapped with siniod or regular coils. Like the coils around the nail to create the magnetism affect when voltage is run through.

7.   The craft can be outfitted with seats etc. also there is another diagram of the consciousness ship that is a cigar shape which has many seats and also four consciousness drive systems.

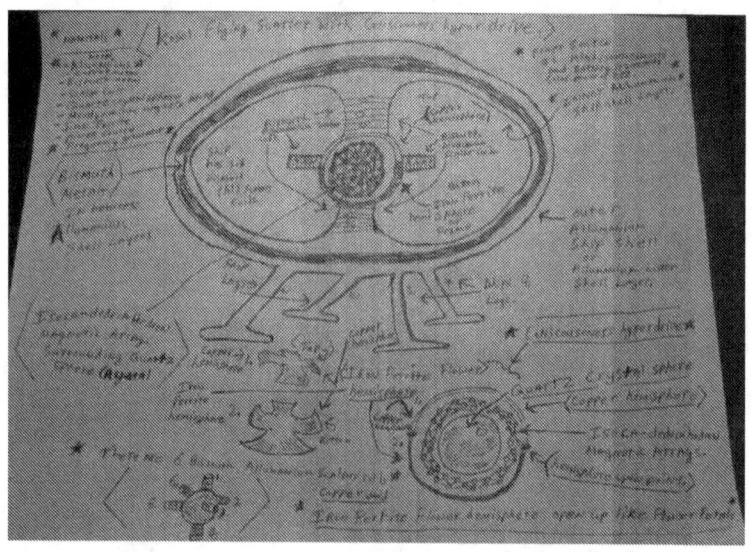

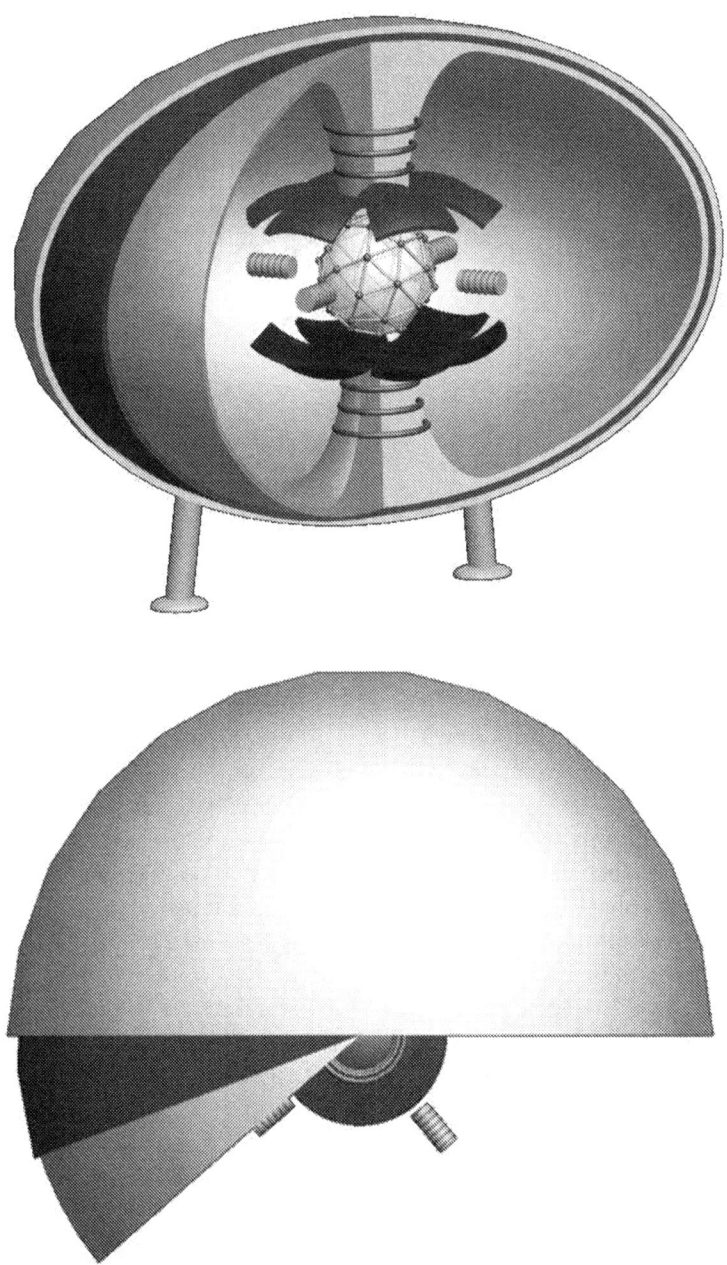

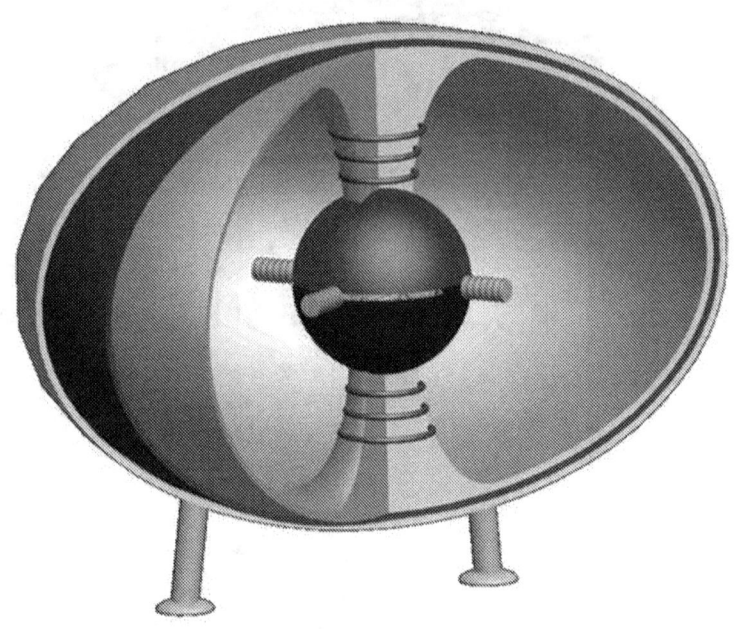

The ship can be outfitted with frequency generators that can produce square and sign wave as well with many amp and voltage power source. The ship is run also on zero point power source. From the scalar coils and siniod or regular coils when together are connected this two coils that is scalar bismuth coil and the siniod coils produce electricity. As well if these two coils is wrapped on a donut shape magnetic either a neodymanium or ferrite. These coils will produce over unity and free energy or zero point energy also. So therefore there is no limit.

This consciousness drive ship is mind and consciousness interactive. They take you anywhere on earth or in the universe just by thought and desire of the craft operator or craft facilitators or pilots. No limits what so ever.

The consciousness drive ship just like the rain maker device can be converted to also to catcher disembodied consciousness units or ghost entities by the desire of the craft operators in the case of the consciousness drive craft but for the RainMaker is north magnetic polarity turning inwards towards the iron ferrite to create the inflows that will catch ghosts.

As for the 3sd and 5sd or five sphere device, static generators (8 of the static generators, van de graph generators etc.) or a tesela coils 8 of them can be added to the five sphere device and three sphere device, to help increase the electrostatic charge fields of both 3sd and 5sd. These electrostatic generators or tesala coils are used for the 5sd and 3sd device to help the device to have electrostatic charge. This electrostatic charge is responsible for the gravitational field and space warp generation. The electrostatic charge created by the electrostatic generators is responsible for the device to have density travels, time travel, hyper space travel and antigravity effect etc.

Because the electrostatic charge is spread throughout the sphere of aluminum or copper metals and the charge then continues to build up and rotate as well as spread throughout the 3sd or 5sd as well the space surrounding the device and its operators and crew. The frequency of the electrostatic is raised in frequency by the rotations of the 3sd or 5sd as well the electrostatic charge can be raised by the frequency generators and scalar coils or siniod coils of the 3sd or 5sd. No limit what so ever.

I stress again the electrostatic generators, Van de Graph generators and the tesala coils is very important in the success of the Kosol 5sd device and the Kosol 3sd device, either one can be used. Now once the frequency of the electrostatic is controlled by the frequency generators or the rotation and counter rotations of the 3sd or 5sd, this will give you access to different dimensional planes of existence to visit and travel to. Remember electrostatic is the key here because electrostatic charge is responsible for gravitational fields generation that is responsible for density plane of travel to other worlds, time travel, teleportation, density travels, other reality, other planet, other galaxy and other density of existence, etc., that includes teleportation of craft and crew, healing, ascension, etc.

Now as you can see the element involved, electrostatic charges, diaelectric like barium tenate, magnetism, spinning, frequency, furthermore David Lowrance will create a bigger aluminum bismuth copper scalar coils for the RainMaker device and will introduce dielectric materials to the quartz sphere and the copper density sphere as well as density sphere. This in turn will create a capacitors like affect that will generated a bunch of electrostatic charges around the RainMaker device as well as creating a lot of electrostatic charges around the copper density sphere and its copper coils also that in turn will create and generate gravitation warps effects, black glowing fogs or black glowing ionizations etc., for density travels, time travel, teleportation, etc., for the RainMaker device and the density copper sphere.

This technique can also be applied to the consciousness drive ships by introducing the dielectrics material to the quartz crystals sphere also don't forget the static generators and high voltage static circuitry can by added to the RainMaker device for super enhanced affect to create weather on demand from clear sky to rain on demand as well to catch ghosts and release them on a massive scale. Inflow vortex catches ghosts, out flow vortex releases ghosts. Since the consciousness drive ship has big scalar bismuth aluminum copper coils already, everything is good to go on that craft platform. No limit what so ever.

Don't forget to check out my other books, Stargate Ascension, Cultivating Inner Force and Reading People Like a Book, Kosol and Koeun Noun Ouch Spherical Generator, The RainMaker Device, etc.

CHI - Broadcaster

The idea to use a compression sphere in direct broadcasting to the sky was born by a test with my Chi-Converter, consisting of only the sphere and the two coiled cones. I saw clouds disappearing as I directed the one end of the converter outside the window.

Depending on the wiring-interconnections of the two cones I detected one cone more radiating outwards than the other. So, the idea was to make a vertical device with cone pointing upwards. As seen in the drawings of the field broadcasters using for agriculture purposes, I decided to build the parts into PVC-tubes of 3" diameter. Each section of the device has now its own tube, and can be combined with PVC adaptor rings to the other sections. Additional radionic-input section and lens-like focusing unit on top are the actual design state.

More units can easily be added for future developments.

See on the cloud pictures, how easily a clearing effect can be obtained in less than one hour, if not much wind is blowing.

The panorama-picture is taken about 2 hours later.

Real nice white clouds around the cloud-free radius of 1 kilometer - weather as I like it!

The whole tube is about 45 cm high. The squares seen on it are the microswitches to connect all the coils together. For the tests, I used the device inside the room. The fields penetrate the roof, if no concrete and steel structure is above the device. Otherwise there will be strong reflections towards inside of the house and it must be used outdoors.

Now, let's look inside the device:

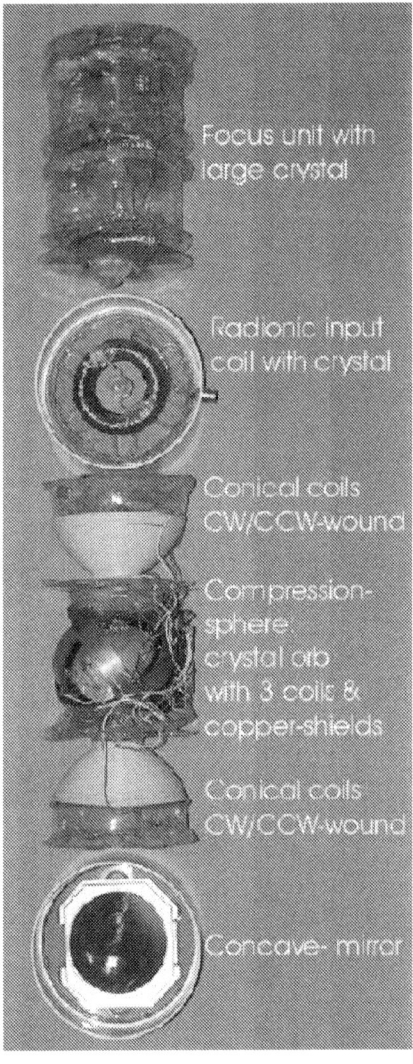

On top of the device is the focus unit with three ring-coils, aligned serially and connected to switcheable capacitors. Inside I glued a large double ended quartz crystal. All coils are wired as shown below.

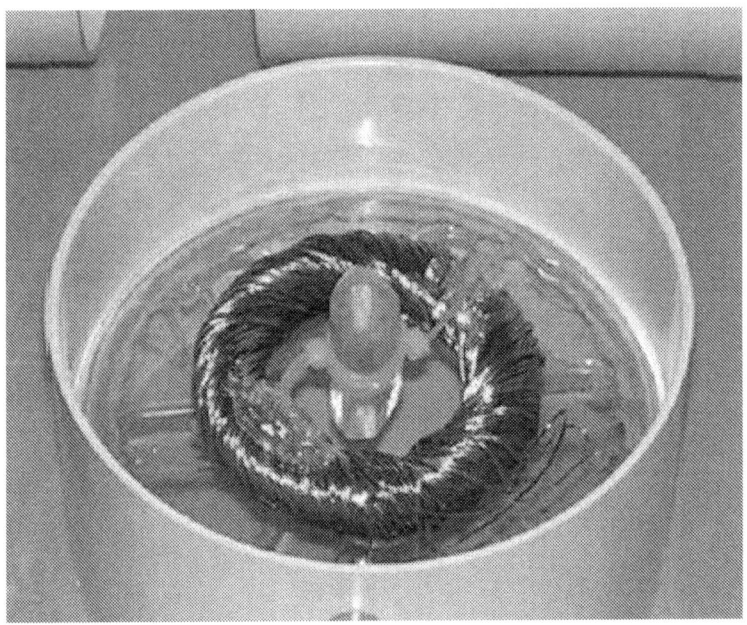

The middle part of the broadcaster is the radionic input section. Again the ring coil with a crystal in the middle. A RCA-jack gives access to the coil from outside. A coiled film box as witness well can be connected here for input of vibrational data. Frequency input or radionic rates via rad-box are possible too.

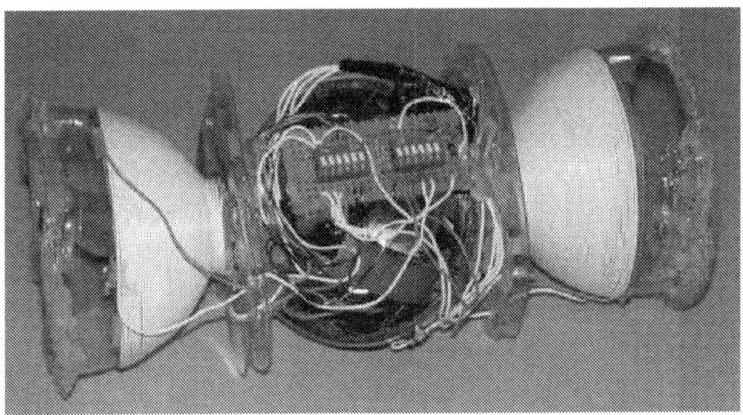

The power section consists of two coiled cones, each having two coils: clockwise and counterclockwise. In between them, the compression sphere is situated. Three coils (using magnet wire) are wound around the crystal orb each 90 deg. angled to the other two coils.

Each coil is covered with heavy copper stripes, connected to the CCW circuit of the cone-coils. Each cone has a double-ended crystal inside.

The switches are used to connect the wirings between the coils. By this all coils can be separated for standby state.

Very important are the capacitor and the germanium diode.

For this first model I used parts of plastic bottles for mounting.

Hot glue was used to keep all together.

Focus- and power unit are designed the way that they can be easily taken out of the tubes for further changes.

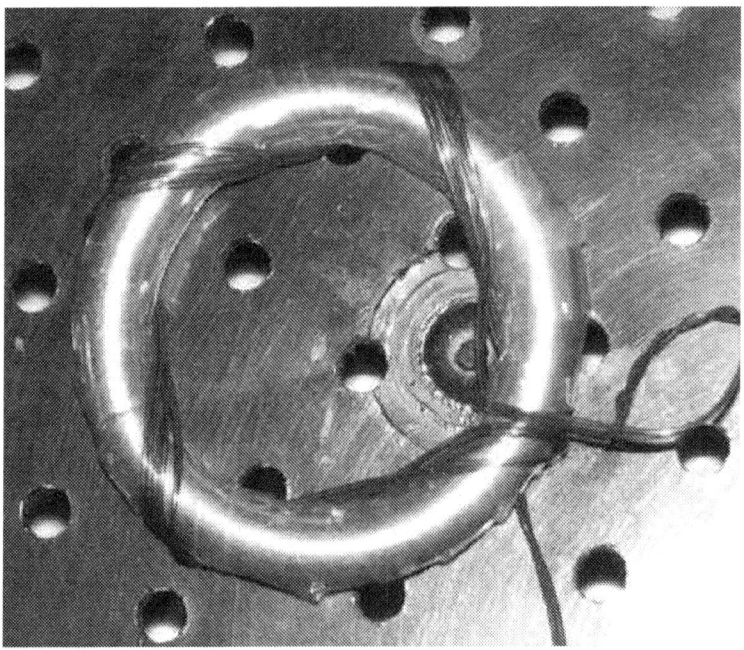

The ring coils all are wound in square winding on a round metal ring with 1/2 wrap before.

First picture shows the first three windings done. Winding starts from upside down inside the ring in clockwise winding with the two open ends of wire fixed as start-point.

Before winding, the ring should be covered with tape.

This shown side on photo is the output-side of the coil and has to face upwards.

The coil finished:

Three of them are for the focusing unit, one is used for the radionic input unit.

The Length of the coils are 27.46 feet plus extra wire for connecting.

Some authors name the length of 20.6" as "Sacred Cubit", other sources say it's the "Royal Cubit", an ancient measure, used in the Great Pyramid too.

The coils use 16x 20.6" as length.

The same length is used on the coils of the sphere and the cones.

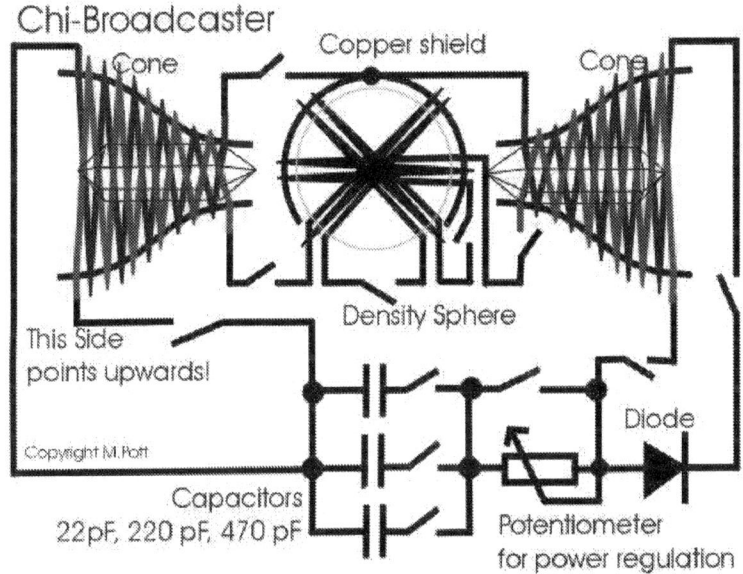

The plan of the power unit:

The cones are wired with two separate sinoidal coilings, one clockwise, one counterclockwise. The clockwise windings are connected to the coils of the density sphere, the counterclockwise windings are connected together and also to the copper shield around the compression coils of the crystal orb.

The lots of switches are necessary to stop broadcasting completely.

At least one of the capacitors must be made active by its switch.

The potentiometer (10KOhm) can regulate the output or can be bridged by switch.

The focus unit has three coils in serial order, connected to three capacitors, each with switch as in the power unit.

This creates a second variable resonance circuit.

Using the concave mirror at the base of the unit, I found an increase of the top output by reflecting the downward directed fields upwards.

The device in this state is a experimental unit with lots of power. One must be careful watch the output and effect on the surrounding area, resp. present persons. If the energy has no room to expand, there can be a quick overload in the surrounding area.

As some people made real interesting experiences with healing-force output of the device using outside in the garden, they could not get enough of it as I observed. In other situations, we had to shut down the device quickly, using as described above inside insulated houses/roofs.

I think this device used as a field broadcaster can give good benefit in agriculture and atmospheric improvements. Using additional homeopathic remedies, real good success is reported with those broadcasters, often used with biodynamical agriculture.

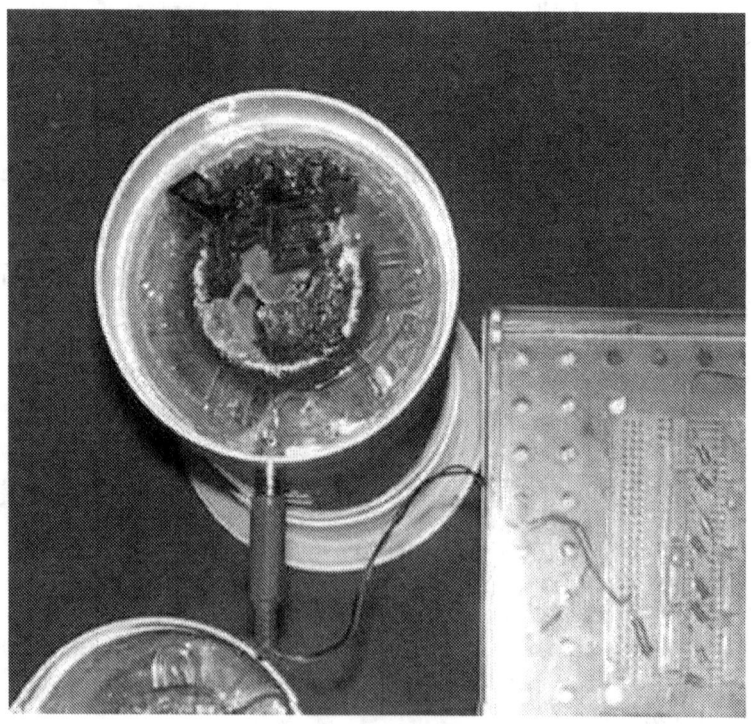

The totally diamagnetic bismuth gives the ability to get direct mental interaction to a vortex resp. torsion wave. It seems to be a doorway to another frame of interaction.

Some bismuth crystals positioned over the input section provide interesting changes. A horizontal small vortex-donut is expanding from the middle of the device, much stronger as the horizontal output of the coils and compression sphere.

This vortex seems to be like a doorway for the mind to go into the bismuth's place, if a person goes to meditative states sitting inside this donut. So here we have a similar situation as well known with the rainmaker.

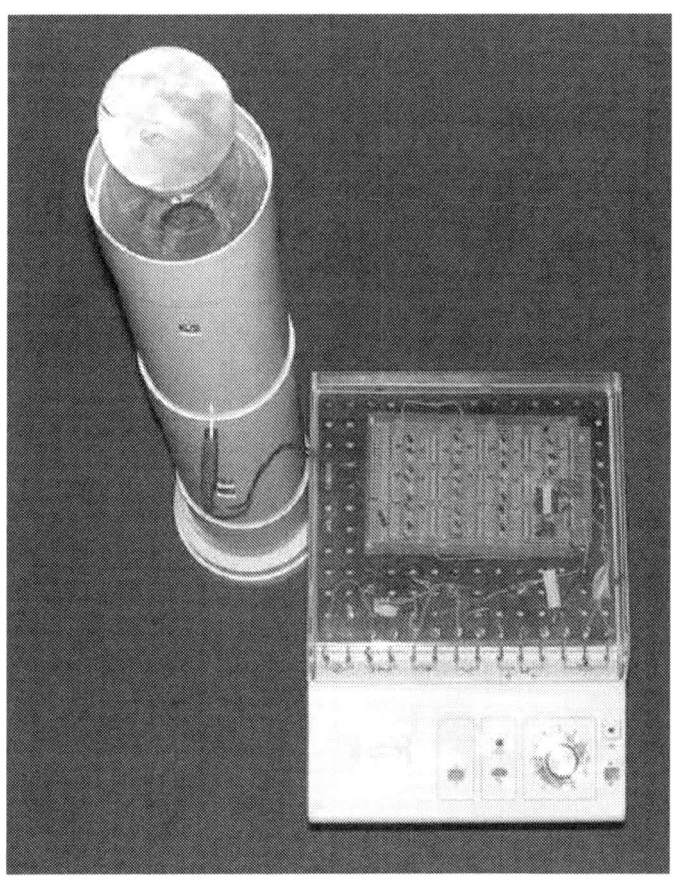

If we use a crystal orb on top of the device, we get a field all around the device instead of the stronger directed beam.

Using a high voltage static generator, a new dimensional change can be recognized. If concentrating on the bismuth donut, the perception is different. It seems like an expanding orb-like space, that draws one's attraction to the outer surface of it, but at the same time there is a connection to the center of the radial movement. This both together creates a meditative feeling being out of time or at the origin of it.

This above are subjective perceptions while doing a little meditation and it's difficult to describe such things with words.

The static generator is a 14x electronic cascade with driver unit using only 9V of batteries to feed the coil with +7000 V DC charge. The coil then is wired with one lead only to the output of the generator.

The output is real stronger as without the voltage.

Light Power Device – Radionics

My new radionic device is ready now.

I went back to totally passive circuits without frequency generators, moebius coils or orgonite layers.

A 3-dial target circuit and a 6-switch-dial trend circuit are in use, dials connected in serial sequence.

For the witness wells I used caduceus coils around to pick up the vibrations of the witness. The caduceus has two different orientations as many other coils, so one should test, which end of the coil is on the bottom or top of the well (pendulum).

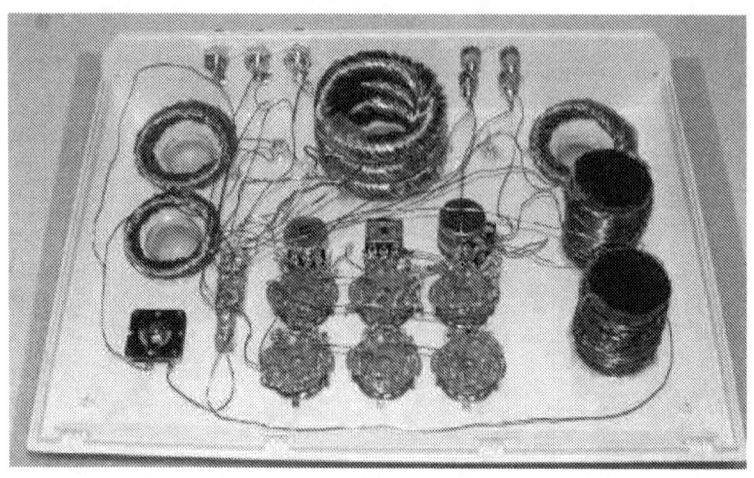

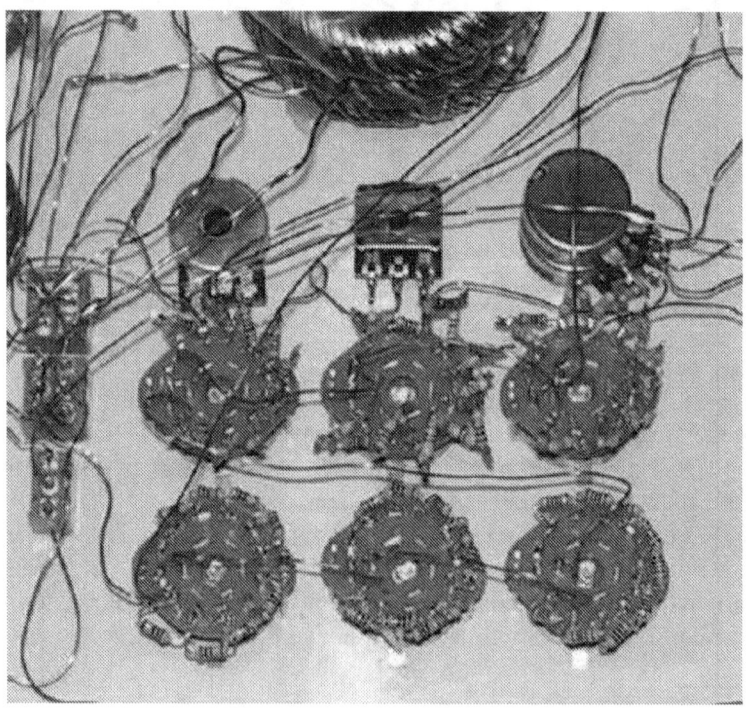

Then I made a third circuit, power circuit, this has variable
capacitor connected to the coils in it.

So now there are three coils in one stack. But where does the power come from? - I used the measure of the "Sacred Cubit" of 20.6" for all the coils: the witness coils: 1x20.6" the output coils 16x20.6".

I can tell you, it's fascinating, how this coils pull the energy from the aether field. I used a rod in like winding on 2" iron rings, but winding it in a square pattern right turning. By this, in each winding we have 2x4 times the 90deg angle cause of additional folding the wire one time before winding - here we get the EM cancelling effect.

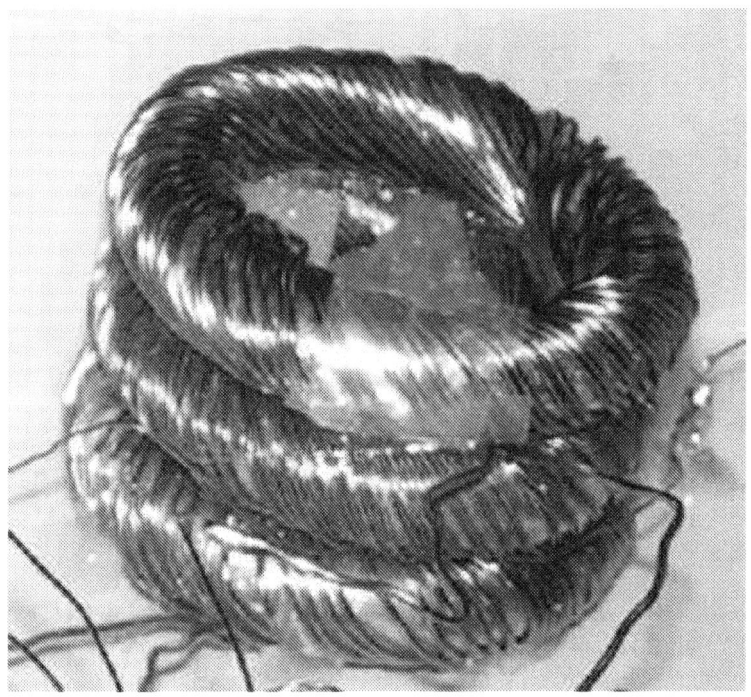

The coils-stack of the three has a crystal inside and on top a crystal orb. Additional Amethyst orb and Crystal pyramid are inserted into the power-circuit with extra ring-coils.

My first tests showed it works. If using no specified target, the orb radiates the vibration into the room. Just inserting a leaf of one of my plants into the target well, the field in the room is gone and goes to the target. Fascinating!

I could not work with the stick-pad. It's too slow to find the results for me. My earlier "ratings" I did with pendulum. Now I feel the according points on the dials inside my own energy system, which doesn't take so much power as with dowsing.

Water Levitation Engine
Designed Kosol Ouch Jan 24 2007

This document now contains several pictures also from a builder
on the Kosol Core Tech forum.

Why don't the clouds fall from the sky?

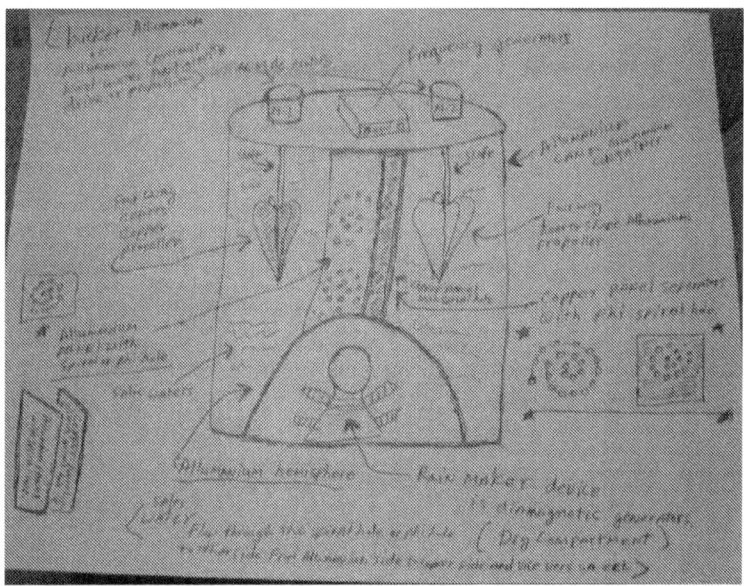

Diamagnetic Generator

RainMaker vortex generator is sitting in the bottom of a large
Aluminum container. At the bottom center RainMaker is setting
with an Aluminum hemisphere over it and glued to the base of the

container. No water gets to the RainMaker unit or on the magnets at all. The wires come out through the hot glue and out the top. The water is setting above and around the RM unit, with the motors spinning near the top of the tank. One propeller is Copper and one is Aluminum and they spin opposite directions. Any diamagnetic generator could probably be used. The RainMaker offers a balanced one with a full regiment of magnets in all directions. The diamagnetic field must travel up the active plates and concentrate there around the holes.

Active Plate

A special active Copper Aluminum plate is installed between the motors propellers to allow the interaction of ions to align the molecules horizontally. The diamagnetic field will propagate along these if they are kept very close and this influence at the active plates is seen to be the heart of the interaction, although the diamagnetic field will also propagate through all the Aluminum container and add to the process of containment. It appears vertically at the holes in the plates. The electric field is spun horizontally into motion and the diamagnetic field is setting on the plates, so molecules setting inside the holes begin to transfer energy from the magnetic field into the diamagnetic field. The holes were shown in a spiral pattern, however this may or may not be symbolic, and would represent the place where the spiral of creation is being accessed.

Plate Distance

Kosol recommends a 1/2 inch gap between plates.

In the tube device experiments it has been shown that the very hot diamagnetic field will move along two tubes at about an inch or less apart for great distances with no observed losses. These can be felt using palming, and one would guess that before the motors are added the distance can be experimented with and felt directly with the hands in the water. The original design was plates touching and the water in the holes becomes highly diamagnetic and water molecules aligned perfectly due to electric ion action.

Water

Sea salt is added to the water, and the heart shaped propeller veins may also be a symbolic shape indicating our link to the water.

Interaction

The copper is charged positive at the atomic layer, and the Aluminum is charged negative, this is due to natural qualities of the elements and also the battery effects of the salt in the water. Each material causes the water molecules to be aligned in reverse because water is also charged. In one side the Hydrogen atoms flip towards Copper in the other, the Oxygen atom will flip towards the Aluminum. The salt acts as an electrolyte to increase this action. It would be assumed that a battery to reinforce the electric field could be added and a bias to adjust for wider plates at the center, but this was not included in the basic design and may effect the spin properties of the salt water restricting its ability to freely spin in the holes of the plates.

As water molecules spin in the holes of the active plates the Diamagnetic field is increased and the water molecules begin to release their torsion coupling between atoms, shrinking the coherent torsion field in the plates and they begin to glow. This is probably a photon release and the first indication that an effect is happening. The diamagnetic field spreads out from the plates to envelope the entire unit. Transmission between water molecules is probably due to the action of the magnetic field accelerating in the water molecules at which time it will begin to spread from the holes and through the water, as the diamagnetic field is aligned with the holes at 90 degrees and will be linking at this point.

It is not truly known where the dual action is located or which way it spins, but I would guess that at the copper plate inside the holes the molecules are spinning the opposite direction of the ones located in the holes of the Aluminum plate. The molecules are probably spinning the opposite direction of the motors as the water is moving in a spiral the holes catch the action and spin in reverse.

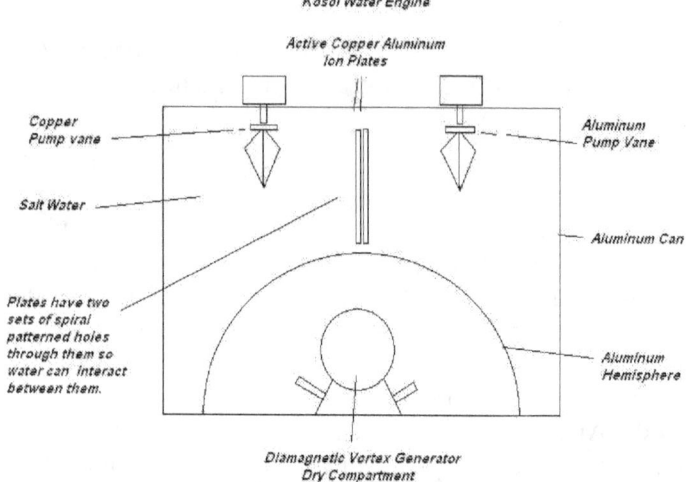

Kosol Water Engine

Active Copper Aluminum
Ion Plates

Copper
Pump vane

Aluminum
Pump Vane

Salt Water

Aluminum Can

Plates have two
sets of spiral
patterned holes
through them so
water can interact
between them.

Aluminum
Hemisphere

Diamagnetic Vortex Generator
Dry Compartment

Kosol Core Tech 1 - 25 - 2007

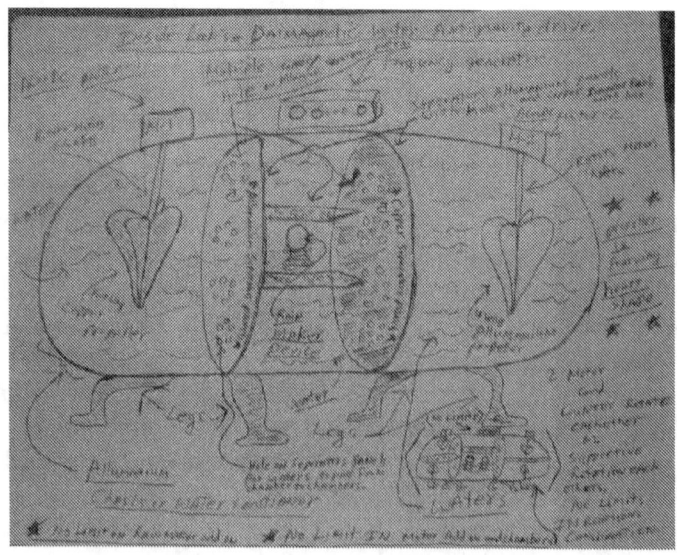

Notes

This diagram shows the correct layout for Copper and Aluminum propellers with respect to inner plates. It will act to give the water molecules a horizontal electric polar alignment, such that they now spin their electric poles and create a magnetic field with both + and - charges in motion. The magnetic field will effect the oxygen and hydrogen protons to flip opposite directions due to each being near the opposite charged side of the water molecule. This will create a diamagnetic field between the Protons and directly lower the torsion field, now on all side of the atom.

Apparently the diamagnetic field from the RainMaker unit is strong enough to cause the interaction in the plates, and the diamagnetic field expands. As the torsion field shrinks the Aluminum and Copper glow blue and red. The outer Aluminum container begins to glow Green. It is though this may be an ionic action but high voltages may be present both inside the salt water and on the Aluminum at various stages of magnetic field collapse.

During the transfer of balance between magnetic and diamagnetic field any electrical coils near the field will begin to conduct Electrons at high rates of velocity, as the waters internal magnetic field is set into motion. The diamagnetic field only responds with a reverse magnetic pressure and no torsion force during the interaction but the now spinning magnetic field will induce electricity until it shrinks. There is danger of electric appliances and lights blowing out if too near the device. It is suggested that a single wire moving through the field would provide an electric current, although what its form will be is yet unknown, probably a DC of sorts, if the device is kept in a balanced state with a magnetic field present.

As the unit becomes diamagnetic, and lowers its coherent torsion connection, at some point it will become possible to repel it totally with a magnet. It will repel from any pole of a strong magnet, and this becomes a method of local control if a total conscious link is not desired, and one can not yet handle the disorientation of disconnecting from the physical body for actual density travel. Magnets arranged on the device in a platonic form, will generate repelling fields in the diamagnetic field as they move towards it. These forces may probably cause motion in the craft.

The good news is that if the devices goes into runaway condition the diamagnetic field will basically decouple it from any possible interactions with the outside world and it may even simply pass through objects, or shoot away at near light speeds to disappear forever from our physical world.

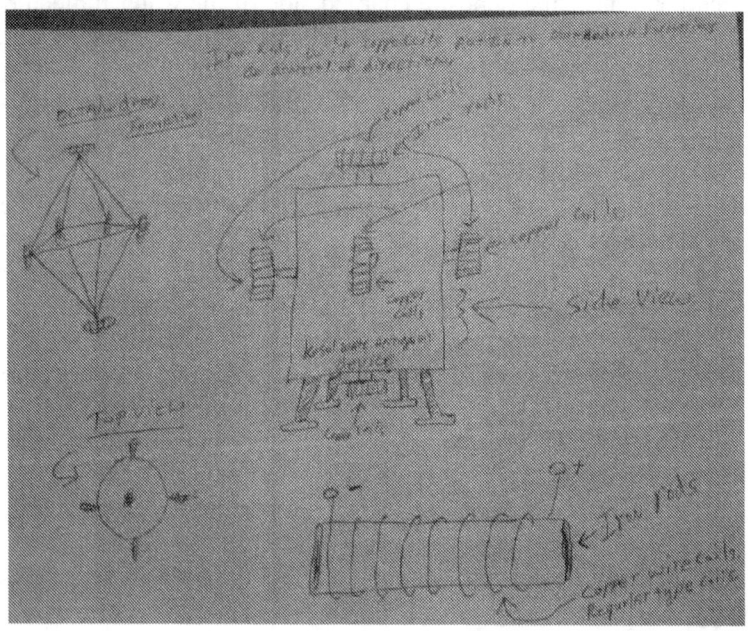

Control may be achieved with external coils, as they energize themselves around a diamagnetic field and then provide the magnetism for repelling the unit.

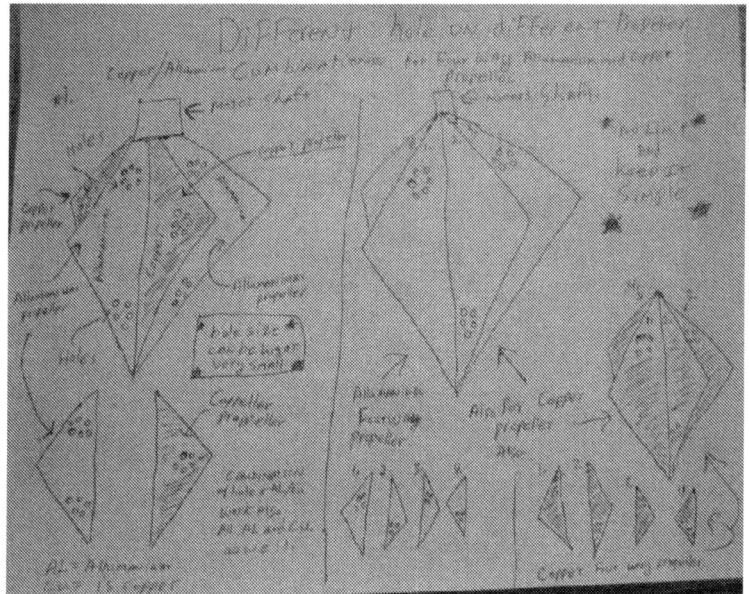

Aluminum and copper propeller combinations and theirs holes combinations for the Kosol anti gravity water engines.

The hole size's can be big or very multiple small. This is true for the separative aluminum and copper panels also.

There is no limit to the metal combinations and size hole as well.

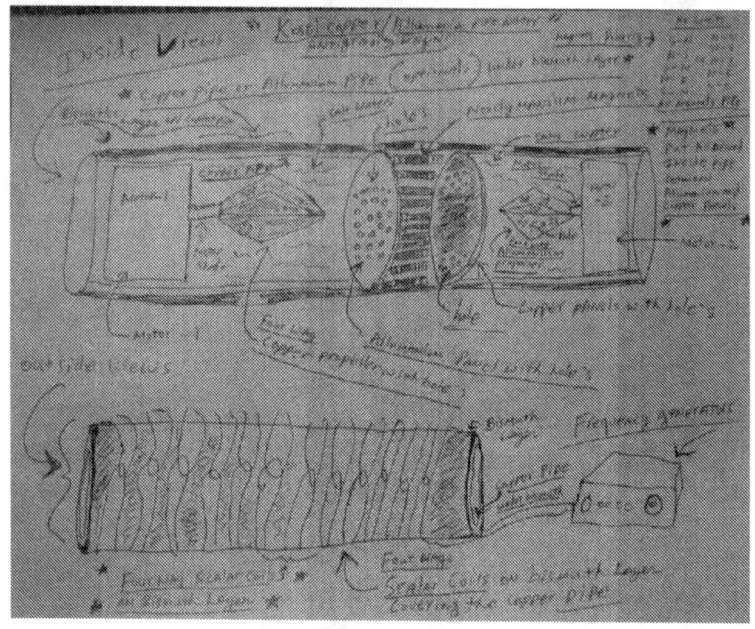

The copper pipe or aluminum pipe is covered with a thins layer of bismuth coatings, and on the bismuth layer is the four way scalar coils.

as well now it has neodymanium magnets on the interior of the device in between the aluminum and copper panel that has many little holes.

the rest is the same like the other version of Kosol water antigravity engine.

Regards, Kosol Ouch.

As Kosol tells me, Water is seen as the source of physical life in the universe and it is considered very valuable, for both humans and energy production.

Here we have two of the most common elements on our planet, Water and Aluminum, combined with small amounts of Copper and sea salt, may be the solution to all our energy needs.

The redeeming thought I would offer for this technology, it will definitely be the Spiritually advanced among us that venture out to become our ambassadors to the other cultures, natures safety valve. But for now, it may be easily possible to contain the effects at a low enough level to create what would appear to be free energy.

On the Diamagnetic Field:

In all my years of researching the atom, early on I recognized that the only force present able to give back more then it takes to operate is the diamagnetic field.

The problem has always been its reverse interacting field. If you send a magnetic pulse into a diamagnetic field you always get back a repulsion and a torsion.

However in our electronics this is always causing a degradation of the signal, and people have searched for the motor that will eliminate this counter EMF, but have never discovered how to use it. I believe the diamagnetic field is emitted from the strong force area of the atom and thus sits where light speed is slightly higher, the neutron. The neutron operates with an overlapped Electron Proton that are as close as they can get to one another. This forms the structure that can react either way to repel EM. Causing an imbalance in the neutron brings up a stronger force, one that is fast enough to regulate the atoms other particles and keep them all repelled into their orbits. The diamagnetic field responds with like fields to repel. This means if you hit it with an Electron magnetic field it responds with an Electron field in repulsion and no torsion is transferred. If you hit it with a Proton magnetic field, it pushes back with Torsion.

Since the diamagnetic field originates at the strong force area and it is normally balanced, it is a neutral field until we interact with it, at which time it will cancel whatever we try to magnetically do to it, only with a faster and stronger interaction because of its higher C.

Dave L

255

John Nelson - Construction Project - Water Engine

John reported his diamagnetic vortex generator, pictured above and ready to be submerged, is the strongest he has built to date, and actually wiped out some of the hard drives on his computer. John has obviously used his own sensory talents to develop a new type of unit for the diamagnetic generator.

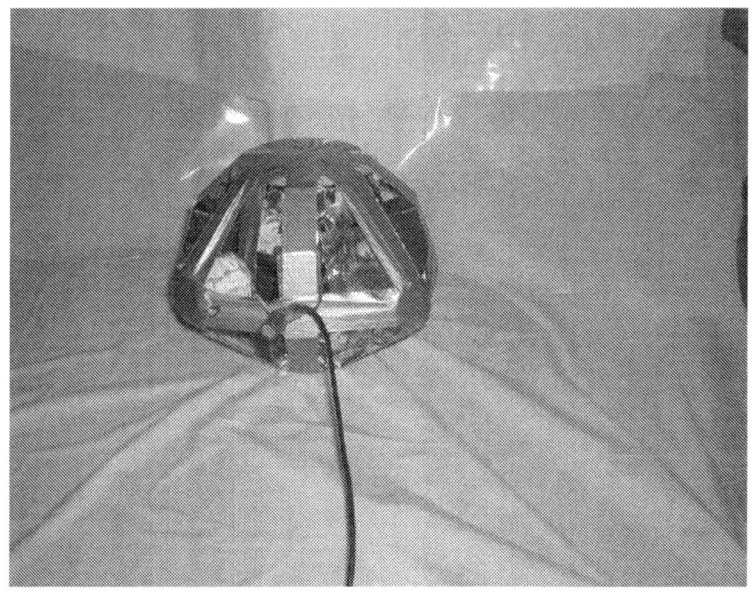

A shot of the internal structure, showing 4 scalar bismuth coils of rather large size and using a capacitor for an added boost.

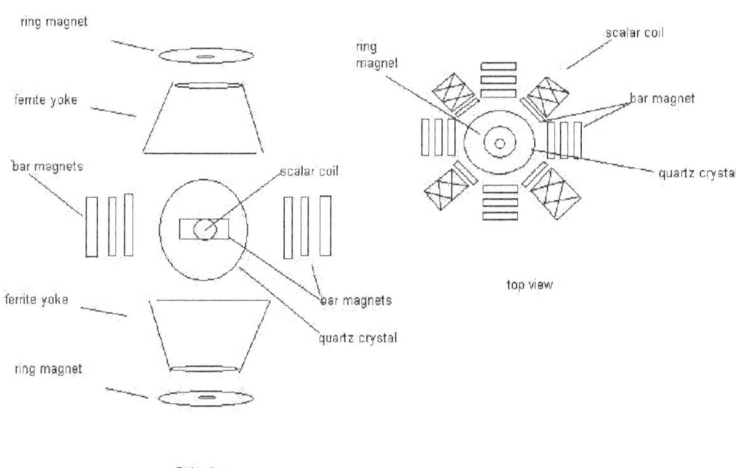

The core is using vertical magnet polarities, and the surface of the Aluminum shell is housing the inwards pointing ones. Two ferrite yoke cores with crystal at the center, and what appears to be another scalar coil inside the ferrites.

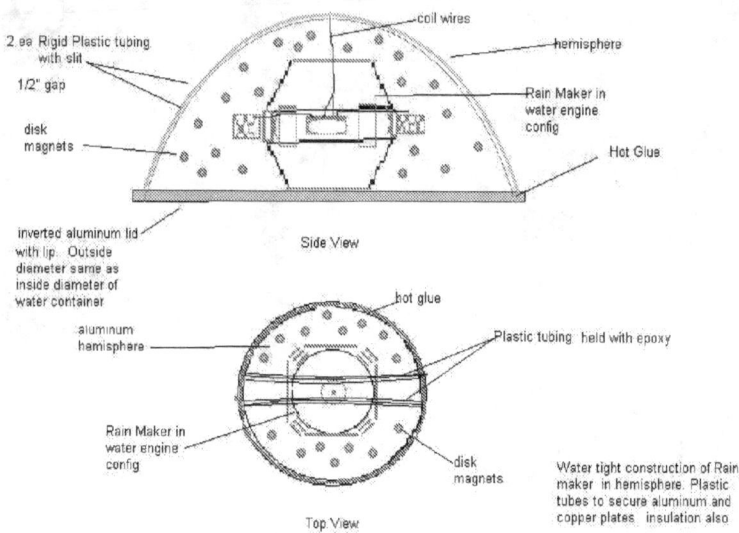

Johns diagram showing detail for the submersible diamagnetic generator. 2 -8 - 2007

Looking down into the water containment tank. It is Aluminum. You can see the plate separator lines that will keep the plates at an even distance.

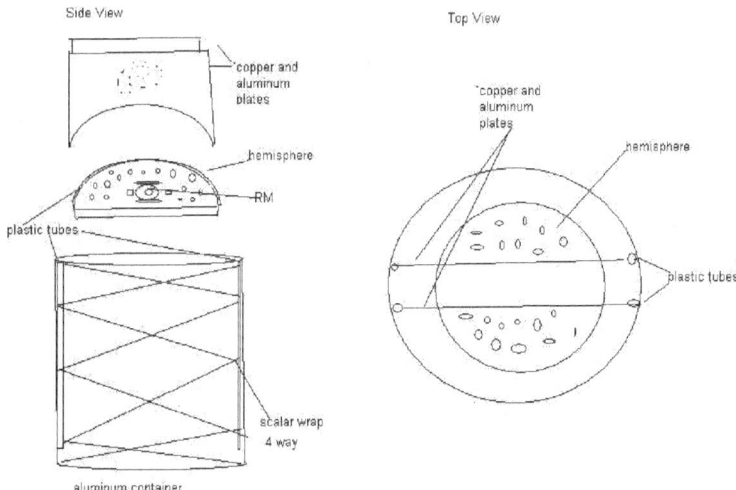

Showing the ion plates above sliding into the main water tank, with the Diamagnetic Generator in the bottom.

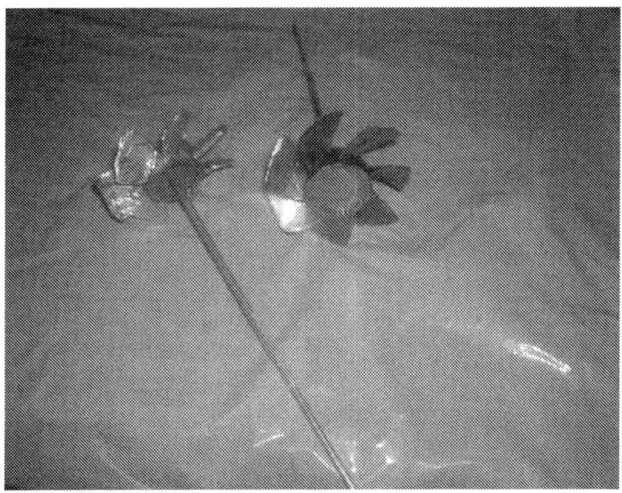

John wrapped PC cooling fans with copper and Aluminum and is driving them with electric drill motors to create the water vortexes. He has reported that the water vortexes form much easier with the diamagnetic generator activated!

Showing the plates with spiral hole pattern.

We thank you John for the excellent photos and diagrams on this project.
Kosol_Core_Tech 2 - 8 - 2007

Kosol states, "It's so simple!"

Fundamental of anti gravity through diamagnetic waters, consciousness and spinning vortex /counter spinning vortex from the guardians.

Dear hearts this is the key, now you understand that there are many ways to accomplished anti gravity.

The whole key was spinning diamagnetic fields, the element that commonly known on your planets is water, water is diamagnetic elements as well also produce diamagnetic fields that can repel both gravity and both poles of magnetism.

When the water element are enhanced with diamagnetic fields generators like the RainMaker 1 and other type of diamagnetic fields generators, then the waters is now given its full potential to be used in practical applications of anti gravity through physical rotations and counter rotations apparatus to spin the highly diamagneticly field charged water elements.

Now dear heart you will notice why our craft stop on lake, pond, oceans, and any waters path way that we encounter, to allow our crafts to suck in water and as we expel the used waters that is pure and put it back into the environment to enhance the life force of that water, and also take in new water to be used and charged with diamagnetic fields.

As you all may know now, there is no limit in our hardness of diamagnetic fields, we can use mechanical version, solid state version, and hydro mechanical version of anti gravity system and propulsion.

Our talent has no limits and our perceptions have no limits, since you all learned from us you also have no limits, like teacher like students.

We are indeed your mentors you are indeed our students and legacy.

As this diamagnetic fields is reach to a highly charged threshold and spinning threshold, the consciousness of the pilot and crew can operate the craft with easy.

I hope now you all understand the power and preciousness of waters, that it create life and is linked to god and to the universe as well the mastery of the star, time travels, nucleus powers as well mastery of eternal life and death is through water, and water is the key to ascension, consciousness and most of all the master of all consciousness and super advance and supreme spiritual technology and physical technology is through mastery of waters.

Regards, Kosol and the Guardians.

Care taker of all universe.

Theoretical and Observations on Water Engine

Water spray shooting out of a high pressure hose is seen to start to fall, and then magically loose its weight slowly drifting upwards. Clouds!

Spinning water molecules may be the key.

Spinning Water

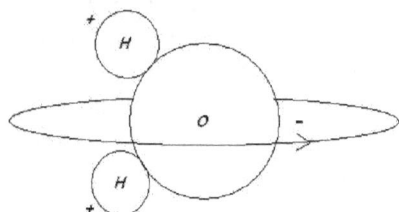

*We see this method involves spinning a positive charge and a negative charge
Thus there is a free standing balance and no DOR should be present.*

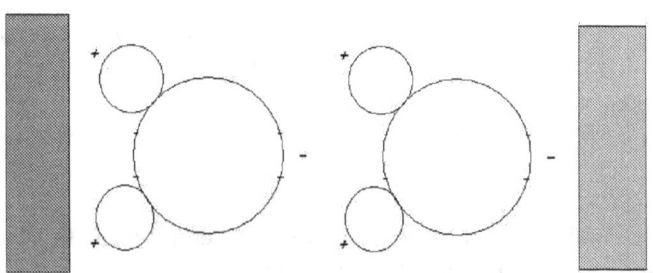

*Aluminum draws Electrons towards it - Copper repells them
This allows us to align the molecules Polarized chains*

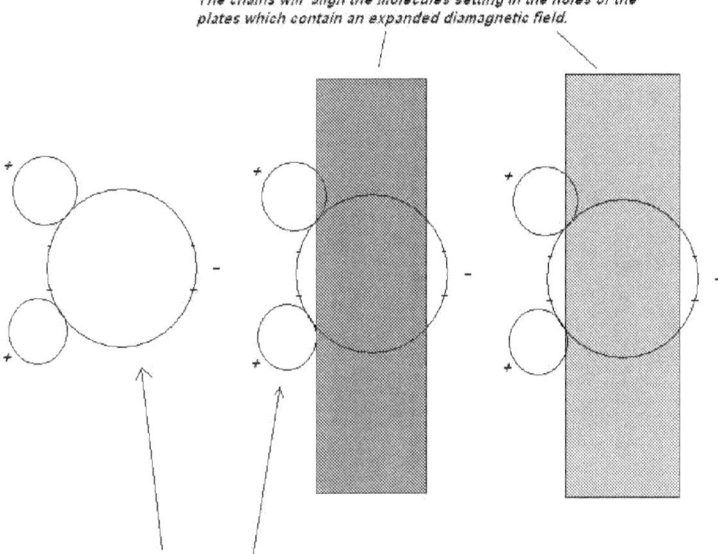

The chains will align the molecules setting in the holes of the plates which contain an expanded diamagnetic field.

Blowing the water along the holes in the plates will spin the molecules. Electric charge attraction.

As long as water pressure is equal on both sides of the plate molecules will not move out of the holes but spin.

The resulting magnetic field generated in the water molecule will meet the diamagnetic field in the plate if present, and cause it to push back harder.

Protons in each atom will align with the magnetic field vertical - and torsion bonding will favor top and bottom of the molecule avoiding the chemical bonds on the molecule.

The Protons will be in reversed polarity and start to create a diamagnetic field from inside. Because charge is reversed in thier magnetic fields - why water is diamagnetic

Kosol Ouch, Koeun Noun Ouch, David Lowrance, Martin Pott, Jerry Evans II and Vince Panella

Water vanes spin opposite directions

Water vane generates a + out charge

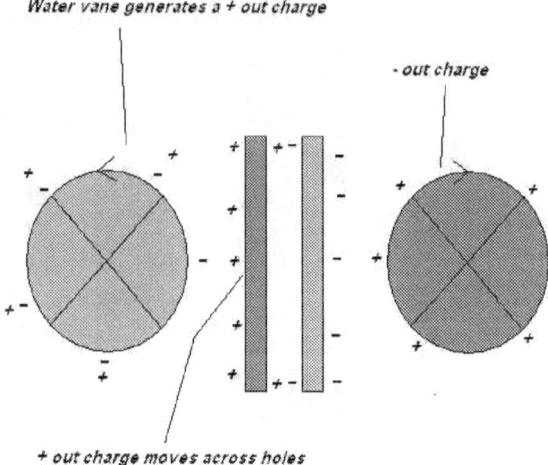

- out charge

+ out charge moves across holes

At some point between the plates the charges reverse their links polarity and a section of water is found able to freely spin, depending on distances of each element.

The active section is probably between the plates and in the holes of the plates where the diamagnetic field begins to expand in the water in the fastest spinning molecules.

The hotest point of the diamagnetic field is setting between the plates.

It is a guess that a scalar coil wrapped on the plates may be enough to activate the water engine.

Another option is to try a copper tube coil and shoot water through it.

264

Diamagnetic Interaction of plates and Water Molecule

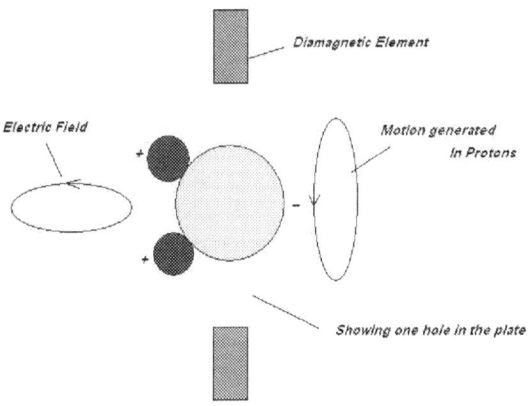

Diamagnetic Element

Electric Field

Motion generated
In Protons

Showing one hole in the plate

Interaction of an electric potential with a diamagnetic substance will produce motion at 90 degrees to the fields path of electrons.

Simply by placing the molecules into a strong external diamagnetic field should lower the spin rate to achieve AG, and possibly allow the water to do this in solution.

This construction sets two diamagnetic fields against one another, the water and the copper. Two diamagnetic substances interacting that counter torsion.

From a standpoint of electrodynamics the water molecule seems unique, because we have access to standing charge as well as the magnetic field of the Protons. But they are freely mobile at the source to spin.

Oxygen   Proton magnetic moment   -2.24077
Hydrogen Proton magnetic moment   4.837353570
          1.21260077   5.159714367   [3 isotopes]

The Protons will align their magnetic fields vertically to the primary spin on the molecule, Hydrogen Proton turned one way and Oxygen turned the other way as Oxygen has a negative magnetic moment. If we could spin them fast enough they would levitate naturally.

The diamagnetic interaction with the copper plates to this emerging magnetic field, from the charge on the electron shell spinning, will be to repel it and create a motion at 90 degrees to the Electric field, a torsion transfer. The major torsion regulation of the atom moves this into the Proton layer from the Neutron seeking to neutralize the action of the magnetic field of the Electron layer and maintain normal orbits, "back EMF".

The torsion from the copper plate will counter the torsion from the waters Nucleus, and the molecule will not spin vertically, however its Protons will if the diamagnetic field is strong enough.

The molecule will want to spin along its magnetic poles, flipping them top to bottom, but it is the Proton spin transferring the torsion, and it will spin first then coupling outwards to the Electron shell. If it succeeds in the torsion link the molecule will start to spin vertically and the induction cycle will become closed into a negative action of back EMF resisting the energy that started it.

However if the diamagnetic field is strong enough outside the molecule effecting the Electron shell to resist this motional torsion, then the Electron shell will continue to spin horizontally and only the Proton shell will spin vertically. Since Copper is highly diamagnetic it will resist the spin with its own diamagnetic field from outside the water molecules and take partial control of the Electron shell. We have succeeded at achieving a 90 degree tilt between Proton spin and Electron spin, only the Proton is now tumbling end over end and can no longer transfer the torsion field outwards of the atom at the same rate as where the magnetic fields were in alignment.

The Torsion field will be reduced and pull inwards, and as we see in clouds this may last a while in the water molecules.

However as the diamagnetic field increases, the effect may spread and critical spin will drop until all the water is enveloped in a diamagnetic field with Protons spinning freely from their Electron shells.

Charge should still hold the water in the container from lifting out, but tempic field density will drop, replaced by diamagnetic fields interacting to cancel one another's effects at the Electron layer. Now setting diamagnetic field against diamagnetic field, both trying to Neutralize the same Electron the diamagnetic field will expand, but with the torsion loop reduced it cannot succeed. Conscious connection can now take control to keep the unit from vanishing.

If the back EMF loop is truly broken in this system, then OU [Overunity] will become a reality as well as AG [Antigravity].

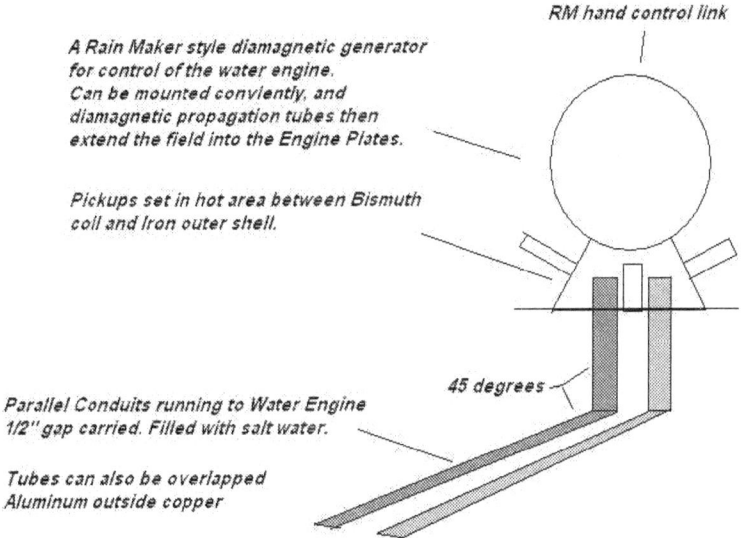

RM hand control link

A Rain Maker style diamagnetic generator for control of the water engine. Can be mounted convliently, and diamagnetic propagation tubes then extend the field into the Engine Plates.

Pickups set in hot area between Bismuth coil and Iron outer shell.

Parallel Conduits running to Water Engine 1/2" gap carried. Filled with salt water.

45 degrees

Tubes can also be overlapped Aluminum outside copper

Theoretical, this system should only be used if you are setting inside the Aluminum structure of a ship. If extended outwards of containment the tubes may become too energized to be safe and arcing may appear between them. It would allow for conscious control through space.

Safety Awareness Bulletins

Acids forming:

There is a possibility that if in some process the water molecules break down with the salts an acid may begin to form. If

267

the water becomes acidic this is an indication of the water loosing its natural chemical bonds and exchanging atoms with the salts. This is not expected to happen, but the possibility exists. A PH tester would be a good thing to pickup.

$H2O + NaCl => HCl + NaOH$ -- not balanced

This is an acid and a base solution so may continually be flipping into the other state

The hydrogen bonding is through Van der Wal's forces

Toxic gases:

Aluminum and salts could also break down to produce some rather nasty gases that would probably want to vent out the top.

Electrolysis:

If the diamagnetic field is not maintained uniformly or strongly enough some of the water may break down to its basic gases due to electrolysis. If hydrogen and oxygen begin to be formed inside a closed container then a possible explosion hazard is present. If this effect is realized then a new free energy hydrogen source has been discovered!

Energy interactions:

When the diamagnetic field expands the area will change its normal qualities. Copper devices will be activated, scalar device will come to life, and magnets may not work the same. If one is inside the field, that is, it is extended all around with an Aluminum shell, electronic devices may not work at all due to the canceling effects of the diamagnetic field. These interactions will appear in steps as the magnetic field drops away and the diamagnetic field comes up in the matter. At full density shift the only connection the outside universe is on the conscious layers.

As well the water engine doesn't have to have water in them the device will work with out water regardless, in this case the water engine will become photonic reactors power source because it has no water in them. As you can see there is no limit.

c_s_s_p group research

Drawings will be updated if there are errors detected.
Dave L
Kosol_Core_Tech

Physical Body Into Light Body

The Guardians said that turning physical body into light body is very easy. The first order of things, is the sun is related to consciousness and the hyperthymus gland which contains the pineal gland and pituitary glands.

Basically you have to look at the sun in the morning and also in the evening.

Absorb the sun light through the eye with breathings and visualizations. Allow the light of the suns from the evening and morning to be collected into the hyperthymus gland which contains the pineal gland and pituitary glands. The pituitary gland and pineal gland once sun light energy can be stored into it can turn the physical body into light body.

Because the hyperthymus gland is responsible for physical ascension and spiritual ascension it can produce soma the ascension hormone.

Ok, here we have the sunlight coming from the sun is absorbed at a certain time which is in the morning and evenings through the eye with breathing and visualizations.

The person must looks at the sun, during that time once he get the full of it, then he or she can just visualizations the entire sun body and see him or herself absorbing the light into the hyperthymus gland in the center of the brain which contains the pineal gland and pituitary glands.

The light is stored there and forms a center of the brain suns which represent the hyperthymus gland which contains the pituitary gland and pineal glands.

With every inhalation the sun light can be absorbed with visualizations or seeing the actual sun, seeing the light coming through the eye with each inhalation and being stored into the center of the brain sun with every exhalation as well you can use mantras along with the whole meditations excising size also.

As time goes by, you will see you body will begin to shrinks from a 5 feet to 3 feet and then you will turn into light.

Now here is another add on, you can use also crystal to harness the sun light. You use a quartz crystal sphere or regular. You can put it on the sun outside and just allow the light from the sun being harnessed with the crystal to be absorbed by your eye with every inhalation and being collected into the center of the brain sun with every exhalation. No limits.

The reasons for the crystal is it helps everyone who is not used to sunlight to have a gentle approach toward this practice as well the crystal helps enhance the frequency vibrations of the sunlight to help create positive affect for those who is beginners. Once you are seasoned you can do without both the crystal and the sun you just use visualizations, but you can use the crystal just to help refine and add on to your auric energy.

Remember once you do this there is no turning back your body will shrink and condense as time go by and then it will turn to a rainbow body or lightbody that will glow physically and will never die. So a 6 feet man will be a 3 feet or 2 feet man. Once he reaches this level then he or she will turn to a light body and you will glow multicolor light radiant and all that.

1. Is posture that full lotus, half lotus, seating on chairs, or laying down.

2. Breathings that slow inhalations through the nose and slow exhalations through the nose. Continue the process of inhale and exhale until you breathing is no longer visible and deep relaxation.

3. The visualizations of the sun as big as you wanted and you draw energy from the sun through your physical eye, it minor chakras in the eyes, and the third eye also or sixth chakras in the 7 chakras system or ninth chakras in the 14 chakras system. Every time you inhale through your nose, and when you exhale your sun energy is collected at the hyperthymus glands which contain the

pituitary gland and pineal glands in a form of a center brain sun. See it all glowing with brightness just like the real sun in the sky. No limit what so ever as well don't forget to put the tongue to touch the upper inner roof of your mouth.

4. As for the quartz sphere crystal or quartz crystal that is used to reflect sunlight so you can gaze your sight on it and as you draw energy from the reflections of the sun light from the crystal with your inhalations and with exhalation you see the energy collected at the center of your brain sun or hyperthymus glands.

You repeat the process over and over until you are satisfied. This meditation can be used to facilitate stargate travel also and RainMaker device unit can be used to help with this kind of meditation as well.

Regards, Kosol Ouch

Ascension Machine and Ghost Catching Device

To tap the zero point energy from the RainMaker device you must have two different metals, for example like an iron bar and copper bar. Each of this metal bars has magnetic bars attached on each of them. But the magnetic bar are separated from the iron or copper metal by a very thin dielectrics such as plastic, paper etc. Then the wire 24 gauge is coiled around this iron or copper metal and magnetic. Now one end of the wire is connected to one end of the copper bar or one end of the iron bar depending which metal you are using. While the other wire end is connected to the device such as light bulb, motors etc. Then the wire end that leads from the other ends of the device is connected to the copper bars metal again at the other ends of the copper or iron bars.

Now you do the same for the other metal bars if you use copper metals, then you can use iron bar metal on the other side of the RainMaker device. Using the same connection as well the wire is coil wrapping on both the iron bar and magnets at the same time just like you did with the copper bars. Connect one lead of wire to the iron and the other lead end to the device and then the device lead wire end to the opposite to the iron bar. Now once the connection is set you move the two different metal bar that has magnetic bar attached to them (the magnetic bar is separated from the metal bar by a thin dielectric materials) close and inside the zero fields or scalar consciousness field of the RainMaker device that the RainMaker device produced around the crystal and around the device itself. Move this two metal bar and magnetic bar that is attaching to it close to the RainMaker crystal about 2 to 3 inches from the crystal sphere.

Then electricity will emerge from the metal bar and begin to run the device that the metal bars are powering. The reason this is possible is because the magnetic fields from the magnetic bars is able to polarize the incoming transitional electrons and incoming transitional protons that is created from the zero fields or scalar consciousness unit energy fields that is created by the RainMaker device and this magnetic field pushes the polarized transitional electrons and transitional proton into the iron bars and the copper bars appropriately. So copper will produce electricity immediately as well as the iron bar also will produce electricity immediately.

Just like the Faraday mono generators, where Faraday has a ferrite magnetic rods and a copper coin that is separated by thin dielectric material. Faraday spun this copper coins and magnet rod that is separated by the thin dielectric material. He connected one wire at the edge of the copper coins and the other wire at the lead light bulbs and the other leads wire is connected from the device to the middle of the coins. Then the electricity was created immediately through this physical spinning, field spins also as the results.

Now with the RainMaker device the device don't spin physically but the zero fields does because of the quartz crystal sphere and the scalars coils as you can see the device now created zero point energy as well it can be tapped by using the described methods of the two metal bar to be used is ideal is copper and iron. Don't forget to wrap both magnetic bar and the two metal bar with a siniod coils as one lead wire end is attached to the metal itself and the other to the device and the lead from the device is attached to the middle of the copper bars or you can attached at the other end of the copper bars to make a complete circuits.

As for increasing the power of the zero point energy from the ZPE coils, density sphere copper coils, and ZPE RainMaker device to 1000 volt and 30 amp of cold electricity not hot but cold electricity meaning the more power you produce the colder the device and electricity it become, the entire circuitry and device must be put into the apex or focus point of the platonic geometry, such as the focus or apex point (inside the middle of the pyramid) of the copper or aluminum pyramids or isocahedron or dedecahedron or sphere or cube or isoca-dedecahedron aluminum or copper platonic geometry, etc., and don't forget to add capacitors (capacitors connected to the ZPE coils and copper density coil), diode (diode is connected to the ZPE and density copper coils if you want DC voltage along with the capacitors) and magnets (magnetic is on the iron ferrite or iron pipe of the RainMaker device) to the circuits as well as the ZPE RainMaker device or frequency generators controlled RainMaker device version.

This is designed to raise human or humanoid collective consciousness, body, mind, environmental, spiritual frequency and frequency so any one who used it, their frequency can be raised to the point where they can become light body as well as glowing or radiant light, so they can interact with light beings in a higher density plane of existence and higher dimensional civilizations and rain or weather affect is the side of affect of the RainMaker device activations. When using the RainMaker device the sensation feels like static charges of hot or even cold around the device, that is how you know it is working, you can feel it with your palm or even your body. Also, the RainMaker device can be used during meditation, waking state, healing session and also sleeping stage no limits what so ever. As well, those who are good with meditation they can use the RainMaker device or don't have too. It is their choice but the RainMaker device does make a good companion for meditations stage, sleeping stage, and waken state, etc. Again, no limit what so ever.

The device called the ascension machine and ghost catching device, see diagram, has quartz sphere wrapped around by density sphere copper wire, then surrounding that is the isoca-dedecahedron magnetic array, then surrounding that is the iron and copper frame, where you can put the four bismuth scalar coils and two siniods coils on to the iron copper sphere frame. The copper/iron sphere frame also is used to hold the magnetic as well. Then the device is put into a octahedron platonic frame or basically any platonic aluminum or copper frame like the cube, sphere, octahedron, isocahedron, tetrahedron, startetrahedron, dedecahedron, isoca-dedecahedron, etc. There is no limit. The coils are connected in series or parallel. Then is connected to the frequency generators or you can just use the coils frequency from the Hatman line or earth grid line that is zero point energy frequency.

No limit what so ever. The frequency generators will give you all range of frequency so you can play with inflow or outflows. As well the magnets are ferrite pick up magnets or neodymanium, that is put around the copper and iron sphere frame that is surrounding the device. As you all know, again, inflow is north polarity of the magnetic turning inwards toward the iron or copper metal sphere frame that is surrounding the device. And out flow is the magnetic north polarity turning outwards away from the iron and copper metal sphere frame is surrounding the device. As you all can see the device is in the apex or focus point of the platonic frame call octahedron that I have chosen for this device.

The device is call the ascension and ghost catching device. Inflow will create rain, portal and will catch ghost or specters and out flow will create healing and ascension physically.

Regards, Kosol Ouch and the Gaurdians Force.

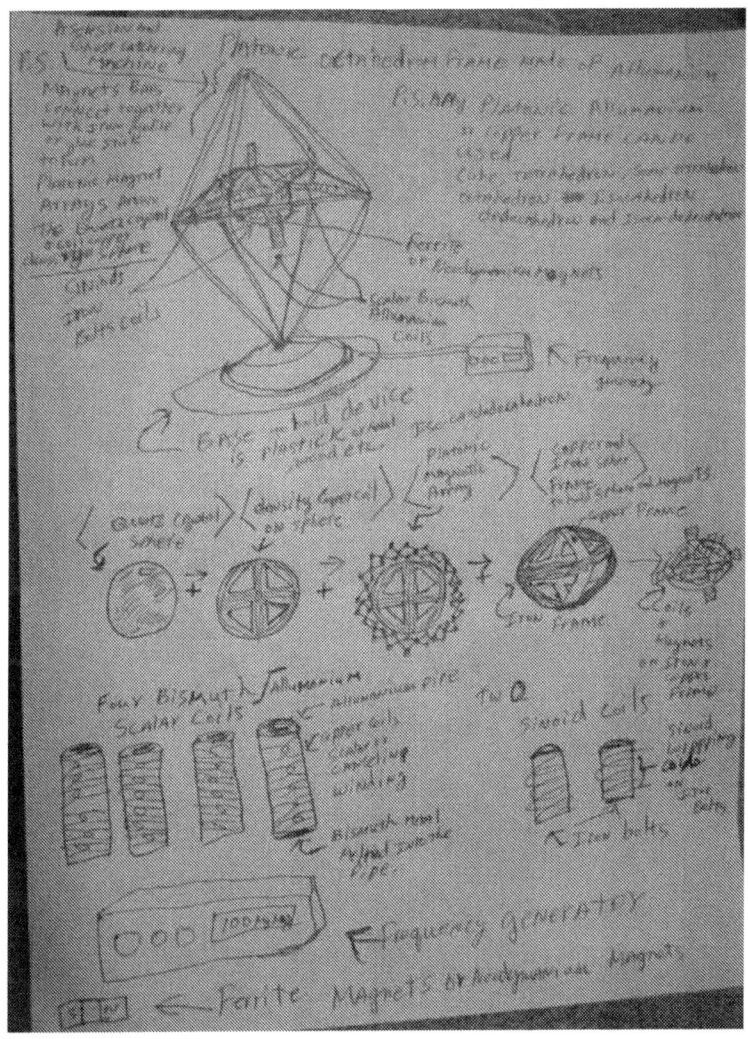

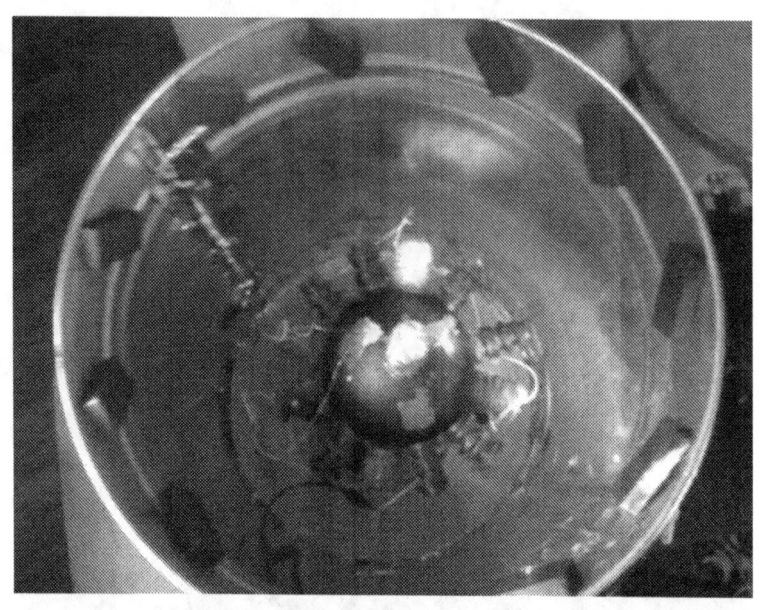

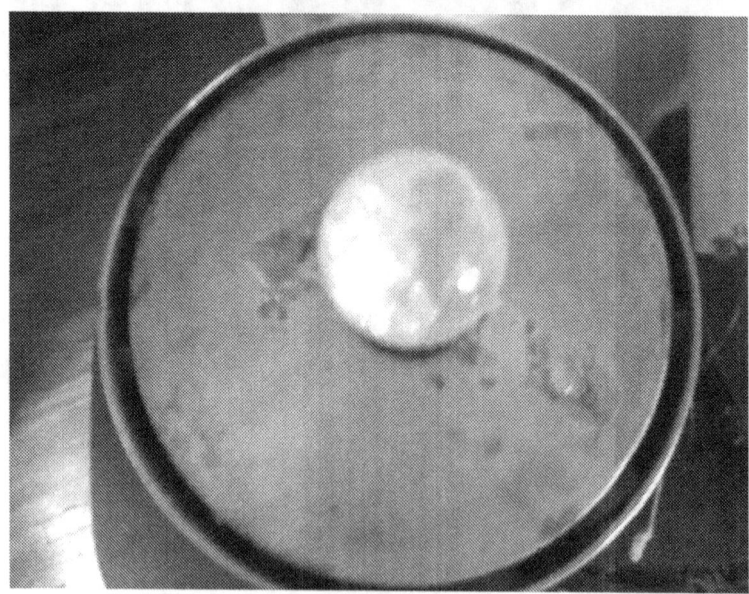

The Gate

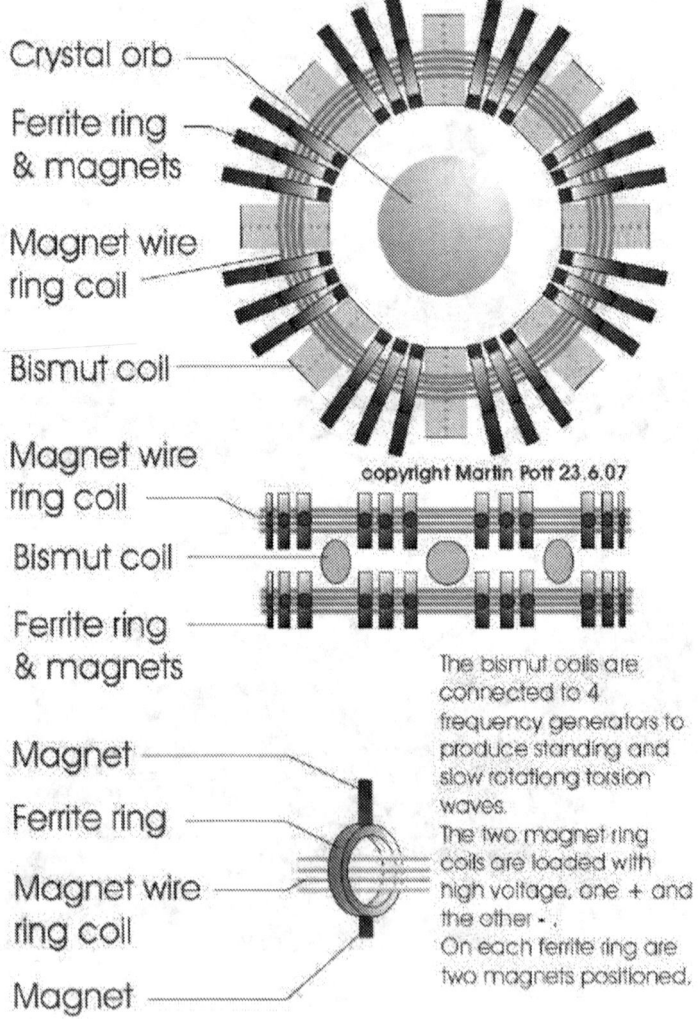

Crystal orb

Ferrite ring & magnets

Magnet wire ring coil

Bismut coil

Magnet wire ring coil

Bismut coil

Ferrite ring & magnets

Magnet

Ferrite ring

Magnet wire ring coil

Magnet

copyright Martin Pott 23.6.07

The bismut coils are connected to 4 frequency generators to produce standing and slow rotationg torsion waves.
The two magnet ring coils are loaded with high voltage, one + and the other - .
On each ferrite ring are two magnets positioned.

Final Comments

The RainMaker can be used as a ghost consciousness and ghost entity catching device when the RainMaker is in the inflow mode that is the magnetic north polarity is all turned inward to create inflow. Now it also depends where you are at on the Earth. Sometime the magnetic north polarity has to be turned outward facing for the vortex to be on an inflow vortex. Again it depends where you are at on the Earth. But most of the time from America to Asia the magnetic north polarity is turned inward facing toward the iron ferrites. So the inflow vortex can now create a consciousness unit that flows inwards and can catch ghost consciousness units. Also the RainMaker device can create vortex of out flow when the magnetic north polarity is pointed outward facing away from the iron ferrite RainMaker base. This out ward vortex will create clear sky and also allow the out flow vortex to create ghost consciousness unit to be released into the environment. As well the out flow vortexs create positive energy healings and will empower anyone who do stargate ascension meditations, both the inflow vortex and out flow vortex will help both the stargate meditations traveler and stargate meditations facilitators to travel through portal of the stargate meditation methods. So is all good all the way round. As well the vortexs both outflow and inflow will help also a healer who do healing on people through rekie method, stargate meditation healing or rechargings method, or even prayer healing or some other meditation methods from yoga budha paths etc. No limit what so ever.

As for the 3sd spherical generators, if the sphere are made from copper sphere all three of them or even the 5sd device then magnet does not need to be added to the sphere. But still the copper coils in 3d platonic formation such as the density sphere

281

method of cooper coils wrapping around the frame of the 3sd or 5sd device. When the 3sd or 5 sphere device are spinning or counter spinning the device will traverse and transverse through different planes of density of reality. So the copper coils is in a 3d platonic formations wrapping around the frame aluminum or steel or iron of the 3sd and 5sd device, and if the copper sphere are used for the 3sd and 6 sphere device etc., then ferrite or neodymanium magnetic are not needed at all on the cooper sphere. The device will work perfectly, no limit what so ever.

Regards, Kosol Ouch

Now below is from Jerry Evans II

My name is Jerry Evans. About 4-5 years ago I began a focused and intensified search for spirituality in my life, and Kosol was a catalyst for that journey. Through the Stargate meditation techniques he taught, I was able to experience periods of calm and deep relaxation, but not the visual and out of body experiences that others were having. Although I had one or two occasions where audio and visual experiences accompanied my personal meditations, I had an overall lack of success. Eventually I struck out on my own journey to learn from others, seek out other healers, and attend seminars in order to further my seemingly slow and unsuccessful search for answers. Over time, accelerated by the teachings and advancements of quantum theory, I began to realize the ultimate and infinite nature of reality. It is completely subjective, and my search for the truth was tainted by my individual perceptions which alone created my own experiences and expectations. By harnessing focused intent, and the desire to heal, I was able to heal others with often instant results. Healings that involved instant physical changes in the human body, and instant energetic changes on the subtler levels of an individual's life patterns. Beyond this realm, time doesn't exist as it's an illusion created by man for the convenience of reference points and measuring tools. Useful as they are, they can be hindering in understanding the nature of how our focused expectations, intent, and perspectives literally create our realities on a moment to moment basis. I found that by focusing on instant results, posing questions to the universe, and exploring different possible outcomes simply by thinking them, that the patient would experience instant changes as I began to explore the infinite possibility of perceiver based interactions with another individual, who's perceptions were also open to change and willing to allow for positive transformation in their lives. By simply desiring a different outcome, and measuring the "before" and "after" of focusing on the desired outcome, along with each individual's desire for healing and change, practically anything was possible. There are no rules about how to go about doing this. Reality is dependent on the individual's perceptions, and we literally create the rules as we go. One use tool I found was simply asking the question. "What if this were different?" "What would it be like, if this were healed, or changed for the better?" This created the platform for a new perception to be perceived, and it would literally manifest as the new paradigm was posed and perceived. So in summary, I've found that learning from others and experiencing various methods of growth, learning, and healing, are all beneficial, it is also important to note that there is not one right way. The world is our playground, so go out there and play the game. Make the rules as you go, and reconnect with the playful child in all of us that seems to get buried as we get burdened with the expectations and mental programming that we are daily bombarded with by media, society, religion, and peers. Let go, and let your intuition be your guide. And most importantly, have fun. This is where I've gotten to at this point in my life, and by all means I'm still learning to harness this new understanding of how I am not just an experiencer, but a creator of all I survey. Life is a journey, and is meant to be fun. Keep this mind, and never forget it. That's what I've come to learn, and this is the platform from which I continue to grow. Most importantly, never discount anything you've experience or learned up to this point. They are, after all, what has gotten you to where you're at. Cherish the past, but don't live in it. Focus on the ever present "now", and create the future you deserve to have.

Jerry Evans II

Also a hello to all my colleagues that work with me in the place of employment in Auburns. That is Michael Brandon Henschell, Nathan Hale III, and Chad Oney who is my very favorite colleague at my place of employment. A special thanks to all of them for their support in welcoming me into the new place of employment in a casino in Auburn.

Regards from Kosol Ouch and all the co-authors.

www.ingramcontent.com/pod-product-compliance
Lightning Source LLC
Chambersburg PA
CBHW072032190526
45165CB00017B/145